INGENUITY IN MATHEMATICS

NEW MATHEMATICAL LIBRARY

PUBLISHED BY

THE MATHEMATICAL ASSOCIATION OF AMERICA

The New Mathematical Library (NML) was begun in 1961 by the School Mathematics Study Group to make available to high school students short expository books on various topics not usually covered in the high school syllabus. In a decade the NML matured into a steadily growing series of some twenty titles of interest not only to the originally intended audience, but to college students and teachers at all levels. Previously published by Random House and L. W. Singer, the NML became a publication series of the Mathematical Association of America (MAA) in 1975. Under the auspices of the MAA the NML will continue to grow and will remain dedicated to its original and expanded purposes.

INGENUITY IN MATHEMATICS

by

Ross Honsberger

University of Waterloo, Canada

23

MATHEMATICAL ASSOCIATION
OF AMERICA

Illustrated by George H. Buehler

Fourth Printing

 Published in Washington by
The Mathematical Association of America

Library of Congress Catalog Card Number: 77-134 351

Complete Set ISBN-0-88385-600-X

Vol. 23 0-88385-623-9

Manufactured in the United States of America

Note to the Reader

This book is one of a series written by professional mathematicians in order to make some important mathematical ideas interesting and understandable to a large audience of high school students and laymen. Most of the volumes in the *New Mathematical Library* cover topics not usually included in the high school curriculum; they vary in difficulty, and, even within a single book, some parts require a greater degree of concentration than others. Thus, while the reader needs little technical knowledge to understand most of these books, he will have to make an intellectual effort.

If the reader has so far encountered mathematics only in classroom work, he should keep in mind that a book on mathematics cannot be read quickly. Nor must he expect to understand all parts of the book on first reading. He should feel free to skip complicated parts and return to them later; often an argument will be clarified by a subsequent remark. On the other hand, sections containing thoroughly familiar material may be read very quickly.

The best way to learn mathematics is to *do* mathematics, and each book includes problems, some of which may require considerable thought. The reader is urged to acquire the habit of reading with paper and pencil in hand; in this way mathematics will become increasingly meaningful to him.

The authors and editorial committee are interested in reactions to the books in this series and hope that readers will write to: Anneli Lax, Editor, New Mathematical Library, NEW YORK UNIVERSITY, THE COURANT INSTITUTE OF MATHEMATICAL SCIENCES, 251 Mercer Street, New York, N. Y. 10012.

The Editors

NEW MATHEMATICAL LIBRARY

1 Numbers: Rational and Irrational *by Ivan Niven*
2 What is Calculus About? *by W. W. Sawyer*
3 An Introduction to Inequalities *by E. F. Beckenbach and R. Bellman*
4 Geometric Inequalities *by N. D. Kazarinoff*
5 The Contest Problem Book I Annual High School Mathematics Examinations 1950–1960. Compiled and with solutions *by Charles T. Salkind*
6 The Lore of Large Numbers *by P. J. Davis*
7 Uses of Infinity *by Leo Zippin*
8 Geometric Transformations I *by I. M. Yaglom, translated by A. Shields*
9 Continued Fractions *by Carl D. Olds*
10 Graphs and Their Uses *by Oystein Ore*
11 } Hungarian Problem Books I and II, Based on the Eötvös
12 } Competitions 1894–1905 and 1906–1928, *translated by E. Rapaport*
13 Episodes from the Early History of Mathematics *by A. Aaboe*
14 Groups and Their Graphs *by I. Grossman and W. Magnus*
15 The Mathematics of Choice *by Ivan Niven*
16 From Pythagoras to Einstein *by K. O. Friedrichs*
17 The Contest Problem Book II Annual High School Mathematics Examinations 1961–1965. Compiled and with solutions *by Charles T. Salkind*
18 First Concepts of Topology *by W. G. Chinn and N. E. Steenrod*
19 Geometry Revisited *by H. S. M. Coxeter and S. L. Greitzer*
20 Invitation to Number Theory *by Oystein Ore*
21 Geometric Transformations II *by I. M. Yaglom, translated by A. Shields*
22 Elementary Cryptanalysis—A Mathematical Approach *by A. Sinkov*
23 Ingenuity in Mathematics *by Ross Honsberger*
24 Geometric Transformations III *by I. M. Yaglom, translated by A. Shenitzer*
25 The Contest Problem Book III Annual High School Mathematics Examinations 1966–1972. Compiled and with solutions *by C. T. Salkind and J. M. Earl*
26 Mathematical Methods in Science *by George Pólya*
27 International Mathematical Olympiads—1959–1977. Compiled and with solutions *by S. L. Greitzer*
28 The Mathematics of Games and Gambling *by Edward W. Packel*
29 The Contest Problem Book IV Annual High School Mathematics Examinations 1973–1982. Compiled and with solutions *by R. A. Artino, A. M. Gaglione and N. Shell*
30 The Role of Mathematics in Science *by M. M. Schiffer and L. Bowden*

Other titles in preparation

Contents

INGENUITY IN MATHEMATICS

Preface

This book contains short, self-contained essays on elementary mathematics. Many describe work of outstanding mathematicians, Gauss among them, and show the incisiveness and ingenuity so characteristic of first rate mathematical work, no matter what the level. I have chosen the topics treated here because I remember how exciting and stirring I found them on first encounter. I want to share this sense of wonder with a larger audience and believe that many people, though unaware of it, have the minimal mathematical background needed to understand and enjoy the ideas described here; the basic material of a standard first high school course in algebra and geometry suffices. It is my earnest hope that some readers will come under the spell of these fascinating topics.

The book is modelled after *The Enjoyment of Mathematics* by Rademacher and Toeplitz, in my opinion one of the best popular mathematical expositions ever written.

The references given in each essay usually include my sources as well as appropriate material for those readers who wish to pursue the topic treated a little farther. The references in the bibliography at the end of this monograph are outstanding books in the general field of elementary mathematics and will nurture a growing interest. In this connection I want to mention specifically the periodical *Scripta Mathematica* which, specially during the years 1935–1957, was a veritable treasure trove of mathematical riches.

In some essays I have treated rather specific concrete problems, often neglecting or merely hinting at generalizations, while in others I have tried to generalize and abstract a bit more. The reader should be aware that the treatment of any particular topic here is just one of several possible approaches. At the end of most essays I have added related exercises whose solutions appear at the end of the book. Since the essays are independent of one another, they may be read in any order.

In conclusion I wish to thank Dr. Harold Shapiro for his excellent appendix to Essay 12 on Complementary Sequences and Dr. Anneli Lax, the Editor of this series of books, for her insight, judgment, patience and kind manner.

ESSAY ONE

Probability and π

I can remember reading years ago that the probability of two positive integers, chosen at random, being relatively prime is $6/\pi^2$. It seems that one R. Chartres, in about 1904, tested this mathematical result experimentally by having each of fifty students write down at random five pairs of positive integers. Out of the 250 pairs thus obtained, he found 154 pairs were relatively prime, giving a probability of 154/250. Calling this $6/x^2$, he found $x = 3.12$, while $\pi = 3.14159\cdots$.

This simply astounded me! How a random choice of pairs of positive integers could have anything to do with π was beyond my imagination. The prospect of actually determining the value of π through an experiment of repeated trials—in which the producer of the pairs of integers has no idea what they are to be used for—seemed utterly incredible.

The mathematics involved in showing the above probability to be $6/\pi^2$ goes beyond the limitations that we have set for ourselves†; however, any bewilderment one might feel about π entering into such results can be resolved by considering the following simple example:

† An account of this matter appears in A. M. Yaglom and I. M. Yaglom, *Challenging Mathematical Problems with Elementary Solutions*, Vol. I, Holden-Day, 1964, San Francisco. Problems 92 and 93 are the appropriate ones.

The reader might also consult G. Pólya, *Mathematics and Plausible Reasoning*, Vol. 1, Princeton University Press, 1954, Princeton, pp. 19ff., on the history of the result $\sum 1/n^2 \rightarrow \pi^2/6$.

The probability that two positive numbers, x and y, both less than 1, written down at random, together with unity, yield a trio of numbers $(x, y, 1)$ which are the sides of an obtuse-angled triangle is $(\pi - 2)/4$.

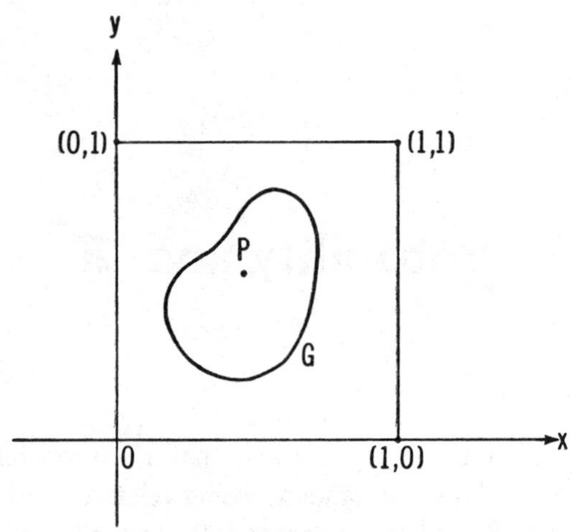

Figure 1.1

First we observe that each pair of numbers x and y determines a point $P(x, y)$ in the unit square (Figure 1.1) whose coordinates are (x, y). Since each coordinate is randomly chosen from the unit interval, the corresponding point $P(x, y)$ has an equal chance of occurring anywhere in the square. More precisely, the probability of P falling within a region G of the square is the ratio of the area of G to the area of the whole square. Since the square has area 1, the probability of P lying in G is the area of G itself.

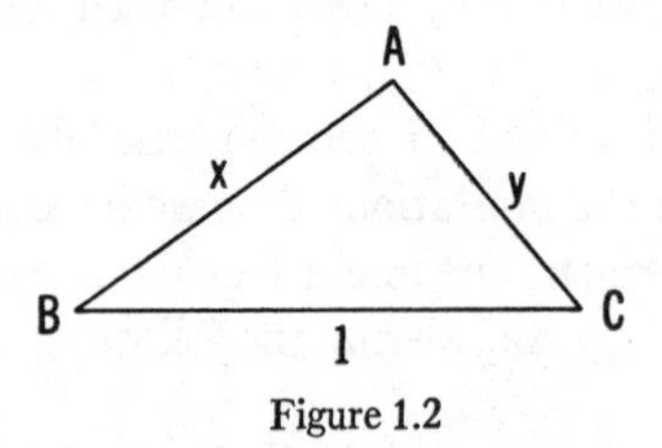

Figure 1.2

Next we consider a triangle with sides x, y, 1 (Figure 1.2). Each of the sides x and y, being less than one, is smaller than the side $BC = 1$. Since the greatest angle of a triangle is opposite the longest side, we see that the angles B and C are less than the angle A. And since only one angle of a triangle can be obtuse, angle A is the obtuse angle in $\triangle ABC$ if such an angle occurs at all.

Now in order for the lengths x, y, 1 to form a triangle of any kind, the sum of any two of them must exceed the third. Clearly the two relations

$$1 + x > y \qquad \text{and} \qquad 1 + y > x$$

hold for all x and y in the interval $(0, 1)$. The requirement that a triangle is formed at all, then, boils down to the condition

$$x + y > 1. \tag{1}$$

If (1) is satisfied, $(x, y, 1)$ will yield a triangle, but it might be acute-angled or right-angled. To see that the type of triangle depends upon the value of $x^2 + y^2$, we apply the law of cosines to $\triangle ABC$ to get

$$1^2 = x^2 + y^2 - 2xy \cdot \cos A, \qquad \text{or} \qquad x^2 + y^2 = 1 + 2xy \cdot \cos A.$$

If A is an obtuse angle, $\cos A$ is negative; otherwise not. Hence the condition that $\triangle ABC$ be obtuse-angled is

$$x^2 + y^2 < 1. \tag{2}$$

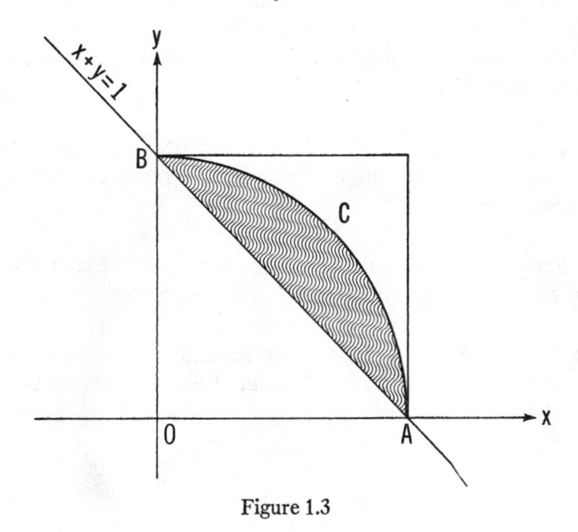

Figure 1.3

Now the points (x, y) satisfying inequality (1) lie above the diagonal AB of our unit square (see Figure 1.3); and the points satisfying inequality (2) lie inside the unit circle. Hence the set of points satisfying both (1) and (2) lie in the shaded region between the quarter-circle and the diagonal, as shown. The probability, then, that $(x, y, 1)$ yields an obtuse-angled triangle is

$$\begin{aligned}\text{area of segment } ABC &= \text{quarter-circle } AOB - \text{triangle } AOB \\ &= \tfrac{1}{4}\pi(1^2) - \tfrac{1}{2}(1)(1) \\ &= \frac{\pi}{4} - \frac{1}{2} \\ &= \frac{\pi - 2}{4}.\end{aligned}$$

EXERCISES

1. Two people agree to meet at a given place between noon and 1 p.m. By agreement, the first to arrive will wait 15 minutes for the second, after which he will leave. What is the probability that the meeting actually takes place if each of them selects his moment of arrival at random during the interval from 12 noon to 1 p.m.?

2. A rod is broken into three pieces; the two break points are chosen at random. What is the probability that the three pieces can be joined at the ends to form a triangle?

3. Three points, A, B, C, are chosen at random on the circumference of a circle. What is the probability that $\triangle ABC$ is acute-angled?

4. A point P is chosen arbitrarily inside an equilateral triangle. Perpendiculars from P to the sides of the triangle meet these sides at points X, Y, Z. What is the probability that a triangle with sides PX, PY, PZ exists?

ESSAY TWO

Odd and Even Numbers

There are some quite interesting problems that require one to show a certain number is either odd or even. Our main concern in this essay is a problem of this kind, but before tackling it, we shall warm up on a couple of easier ones.

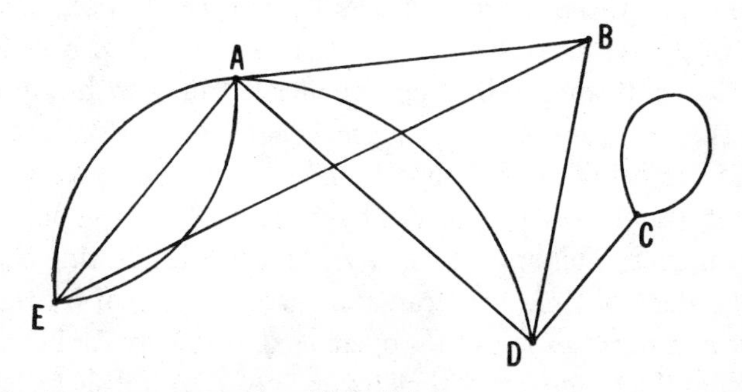

Figure 2.1

(i) Let a finite number of points A, B, C, $\cdots$, K be joined in pairs by any system of curves, including the possibility of loops (e.g., joining a point C with itself; see Figure 2.1) and of multiple edges (joining the same pair of points). We define the *local degree* of a vertex A, denoted by $d(A)$, to be the number of edges incident with the point A, counting loops twice. For example, in Figure 2.1,

$$d(A) = 6, \quad d(B) = 3, \quad d(C) = 3, \quad \text{and so on.}$$

PROBLEM: Show that in any network, as outlined above, the number of vertices which have odd local degree is an even number. (Note that in the example above, precisely *two* vertices, B and C, have odd local degrees.)

The proof in general is simple. We denote by T the total of all the local degrees:

$$(1) \qquad T = d(A) + d(B) + d(C) + \cdots + d(K).$$

In evaluating T we count the number of edges running into A, the number into B, etc., and add. Because each edge has two ends, T is simply twice the number of edges; hence T is *even*.

Now the values $d(P)$ on the right-hand side of (1) which are even add up to a sub-total which also is even. The remaining values $d(P)$ each of which is odd, must also add up to an even sub-total (since T is even). This shows that there is an even number of odd $d(P)$'s (it takes an even number of odd numbers to give an even sum). Thus there must be an even number of vertices with odd local degree.

If we think of the vertices A, B, C, $\cdots$, K as people, and the joining of two vertices A and B (say) to mean that A and B shook hands (loops, if any, indicating one shook hands with himself and counting as two handshakes), the local degree $d(A)$ of vertex A gives the total number of times A shook hands. The above result, then, tells us that the number of people who have shaken hands an odd number of times is even. This application is all the more interesting because it is independent of time—one can state without fear of contradiction that the number of people at the opera next Thursday (or in the whole world from the beginning of time if you like) who will shake hands an odd number of times is even. (One might enjoy verifying this result with a group of friends.)

(ii) Let a_1, a_2, a_3, $\cdots$, a_n be any permutation (arrangement) of the numbers $1, 2, 3, \cdots, n$. Form the product

$$P = (a_1 - 1)(a_2 - 2)(a_3 - 3)\cdots(a_n - n).$$

PROBLEM: If n is odd, prove that P is even.

The solution is very easy. If n is odd, the sequence $\{1, 2, 3, \cdots, n\}$ begins and ends with an odd number, and so contains one more odd number than even numbers. Now the numbers $a_1, a_2, \cdots, a_n$ are these same numbers $\{1, 2, \cdots, n\}$, in some other order. As a result, two odd numbers must wind up together in the same bracket $(a_r - r)$. We see this by noting that an attempt to put both an odd and an even number together in every bracket is doomed to failure because there are not enough even numbers in the collection $\{1, 2, \cdots, n\}$ to be paired with all the odd ones; running out of even numbers forces two odd ones together. Any factor in which this occurs is an even number (the difference between two odd numbers), making the whole product P an even number.

(iii) In this problem, we take any triangle ABC and partition it into smaller triangles in any way whatsoever that does not put a vertex of one of the new triangles on a side of another new triangle. (See Figure 2.3; the situation in Figure 2.2 is the only one not permitted.)

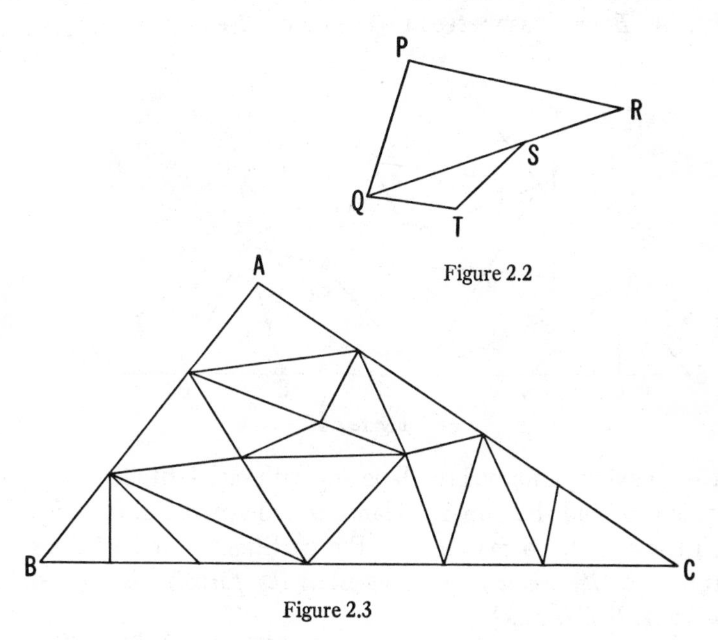

Figure 2.2

Figure 2.3

For example, we might place any number of points into the interior and on the sides of the given triangle ABC and then begin joining those pairs of points that can be joined by segments which neither go through another one of our points, nor intersect segments already drawn in this manner (except, of course, at their endpoints). When we have joined all pairs of points that may be joined in this manner, we shall say that the original triangle has been fully partitioned. This method affords us great freedom, not only in choosing the vertices of the partition, but also in the order in which we connect them; different ways of connecting pairs of points yield, in general, different partitions. The triangles of the partition are all those containing none of our points in their interior, nor on their sides (except, of course, at their own vertices).

Now we encounter even more freedom. On the side AB of the given triangle there are vertices of triangles belonging to the partition. We are to label each of these vertices on AB either A or B, as we please. Similarly, any such vertices on BC we are to label either B or C, as we please; and any on AC either A or C, as we please. At this point, all the unlabelled vertices of the partition occur inside $\triangle ABC$; here we encounter our final freedom—we may label these interior vertices either A, B or C as we see fit. (See Figure 2.4.)

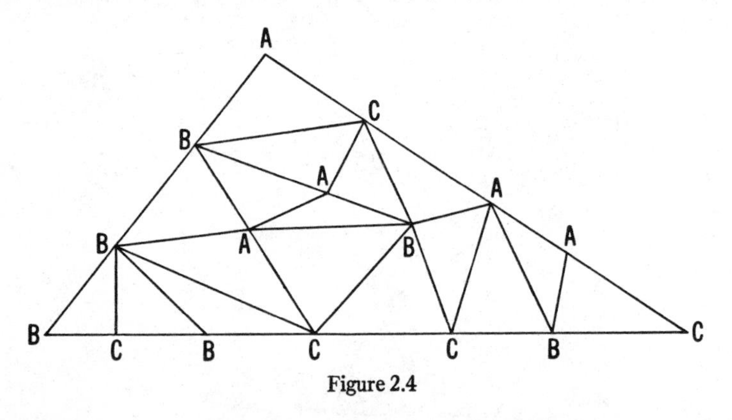

Figure 2.4

Now, having performed this labelling with such latitude, we may be surprised to find that any semblance of order remains. However, we shall show that, no matter how the partitioning and labelling are carried out, *the number of triangles in the partition having vertices labelled A, B, C is odd.*†

† This assertion, also known as Sperner's Lemma, is discussed in L. A. Lyusternik, *Convex Figures and Polyhedra*, Dover, 1963, pp. 160–168, where some of its consequences and generalizations to n dimensions are given.

We distinguish two types of segments: those labelled

(i) AA, BB, CC, with the same letter at each end,

(ii) AB, BC, CA, with different letters at each end.

The former we label 0, the latter 1. We write the sum of the labels of the sides of each triangle inside the triangle. In this way we distinguish three kinds of triangles (see Figures 2.5 and 2.6):

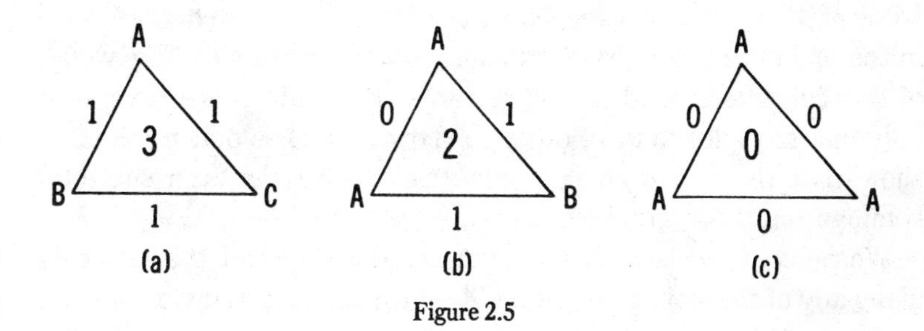

Figure 2.5

(a) triangles with all vertices different;

(b) those with exactly two vertices alike;

(c) those with all vertices alike.

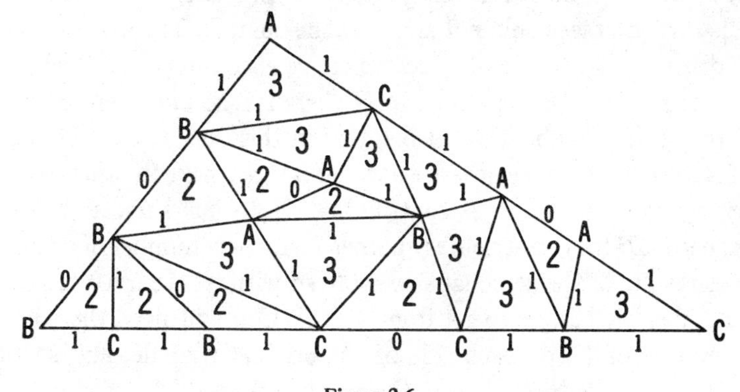

Figure 2.6

We shall prove shortly that the overall sum M of the values of all the triangles in the partition is always an odd number. Accepting this for the moment, we easily obtain our conclusion. The triangles of types (b) and (c), above, contribute even numbers (2 and 0, respectively) toward the value of M, so the sum of their values is also even. Since M is odd, the remaining contributions must have an odd sum. Each triangle remaining is of type (a), and counts 3 toward the value of M. There must, therefore, be an odd number of them. That is, the number of triangles labelled ABC is odd.

Now let us show that M is always odd. First we observe that every segment *inside* the original triangle yields an even contribution to the value of M because its assigned value of 0 or 1 is counted twice, once in the sum of each of the two triangles of which it is a side. The values of the segments around the outside are only counted once (each is in only one triangle), and in order to show that M is odd we need to show that they have an odd sum (to go with the even sub-total coming from the internal segments).

We proceed to show that the sum of the values of the segments along any of the sides AB, BC, CA of the original triangle is an odd number. Since there are three such sides, their grand total will thus be odd. It doesn't matter which side we take; the argument is exactly the same in every case.

Let us consider the side AB, say. At one end is A, at the other B, and an unknown mixture of A's and B's at the (unknown number of) vertices between them.

As we traverse the side starting at A, we encounter the contribution 1 only on sub-segments with differently labelled endpoints. There must be at least one of this kind, otherwise all subsegments would be labelled AA and we could not arrive at the point B. The first segment with distinctly labelled endpoints which we encounter is labelled AB, in this order, since we began at A. If there is a second segment contributing 1, it must be labelled BA, and in this case, there must be a third, labelled AB, since otherwise we cannot reach B. In other words, every change from A to B contributes a 1 to our sum, and every change from B to A contributes another 1. An even number of changes always results in the same label we started with, so we must have an odd number of changes to get from A to B. Consequently, there is an odd number of 1's on each side of the original triangle, and so the contribution of its three sides to the number M is odd.

ESSAY THREE

Sylvester's Problem of Collinear Triads

In connection with the geometry of cubic curves, the English mathematician James Joseph Sylvester (1814–1897) sought a finite set S of distinct points in the plane, not all in a straight line, possessing the property that the straight line joining any two points of S contains at least one more point of S. The property is possessed by some pairs of points in the set $S = \{A, B, C, \cdots, I\}$ shown in Figure 3.1. (G, H, I are in a straight line by Pappus' theorem†). However, there are several other pairs, for example A, D or D, I, such that the lines through them contain no other points of S. Of course, it is conceivable that an unsuccessful attempt to construct such a set of points might be made successful by adding a few strategically placed points. In the attempt of Figure 3.1, the point J, the intersection of AD and GH, might be included in the set in order to correct for the deficiency in the line AD. But, with J added, four new deficient pairs are formed (JB, JC, JE, JF).

Given a finite set of points, not all collinear, let us connect each pair of them by a straight line. If a line contains exactly two points of the set we shall call it an ordinary line. Sylvester was looking for a

† Pappus' theorem asserts: If the six vertices of a plane hexagon lie alternately on two lines, the three pairs of opposite sides meet in collinear points.

finite pointset containing no ordinary lines; he was unable to find such a set. Accordingly, in 1893, he proposed to the mathematical world the task of settling whether or not such a set of points is possible. The question was neglected, apparently, until 1933 when T. Gallai (alias Grünwald) showed that for any finite set of distinct points in the plane, not all collinear, there exists an ordinary line. Consequently, Sylvester's set does not exist. Gallai's proof, however, is very complicated. But, as often happens, soon after his success other, simpler proofs were found. The following solution is due to L. M. Kelly.

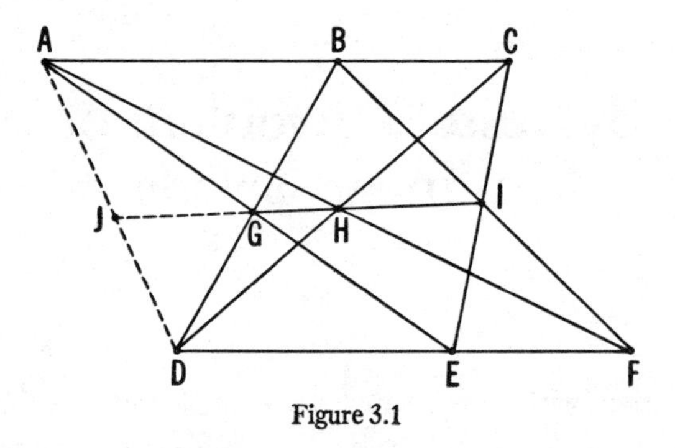

Figure 3.1

Consider any finite set S of points in the plane, not all collinear. Then, any line joining a pair of them does not contain all the points (there must be at least one point not on the line). Consequently, a list of "line-point combinations" may be compiled in which each line l joining two of the points is associated with a point P not lying on it. Since the number of points in S is finite, the number of line-point combinations (l, P), though perhaps very large, is finite also. Therefore, theoretically, a complete list of the combinations may be made and inspected.

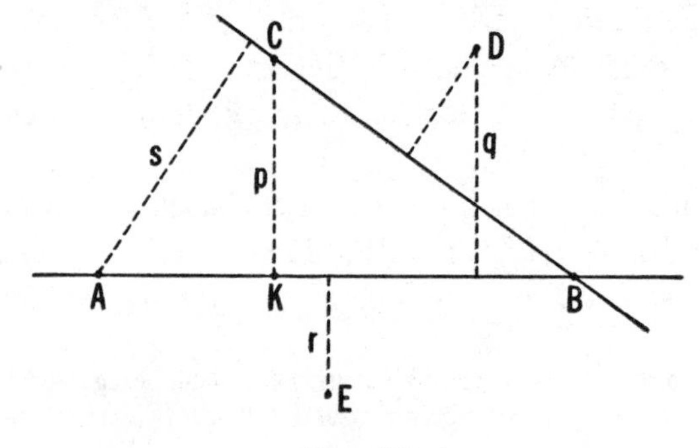

Figure 3.2

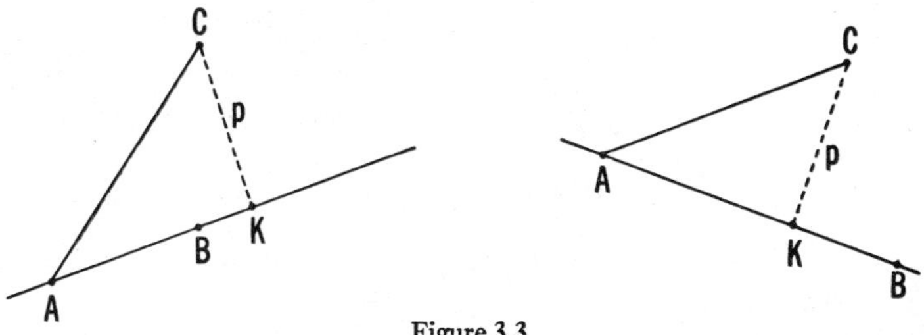

Figure 3.3

With each of these combinations (l, P) we associate the *perpendicular distance* from P to l; for example, with $AB - C$ we associate the distance p from C to the line AB (Figure 3.2). Since the list is finite, there must be a distance which is minimal, i.e., one for which there is none smaller in the list. (If several combinations have this minimal distance, then any one of them will do.) Let the combination $AB - C$ have minimal distance p, and let K be the foot of the perpendicular from C to AB (Figure 3.3).

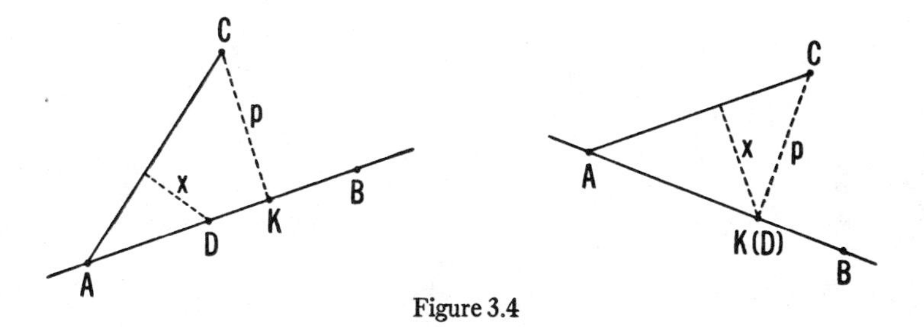

Figure 3.4

We claim that the line l through A and B is ordinary. For, if l contained another point D of the set, then two of the three points A, B, D would lie on the same side of K (or one might coincide with K; see Figure 3.4). Suppose for definiteness that A and D are on the same side of K. Then the combination $AC - D$ has an associated distance x smaller than p. This contradicts the definition of p. Hence the line of a line-point combination associated with the minimal distance is ordinary.

Of course, there may be more than one ordinary line. Kelly and Moser proved that at least $3n/7$ such lines occur for a set of n points. Figure 3.5 shows that we cannot count on any more than this number—exactly 3 of the lines here are ordinary, and there are 7 points. Any triangle shows that *all* the lines may be ordinary.

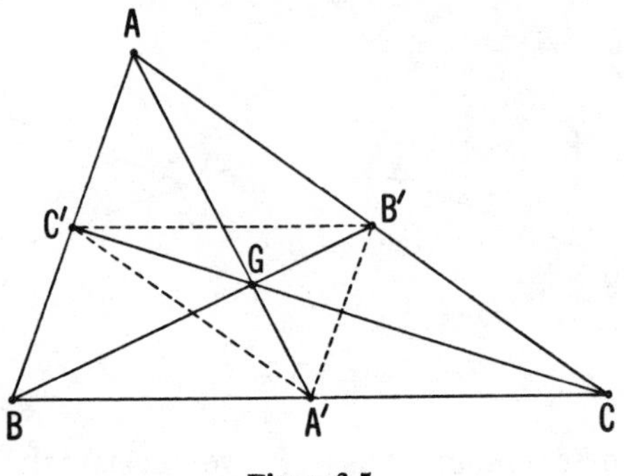

Figure 3.5

It is not at all necessary to use the notion of distance in a proof of Sylvester's problem. In his *Introduction to Geometry*, Professor Coxeter says that to do so "is like using a sledge hammer to crack an almond"; he shows how to establish the result from a meagre set of axioms centred around the idea of "betweenness".†

† H. S. M. Coxeter, *Introduction to Geometry*, John Wiley & Sons, Inc., 1961, New York.

ESSAY FOUR

The Algebra of Statements

In ordinary algebra letters are used to represent numbers; in geometry they represent points, lengths, and so on. There is no reason why we cannot use them to stand for statements (that is, assertions, not questions or commands). For instance, we might assign the letter p to represent the statement

"2 is an even integer".

In the "algebra" of statements, we will use the symbol $\sim$ to negate a statement. Accordingly, $\sim p$ represents

"2 is *not* an even integer".

Suppose, in addition to the letter p above, we let q represent the statement

"π is rational".

We use the symbol $p \cdot q$ (as if p and q were multiplied) to stand for the compound statement

"2 is an even integer AND π is rational",

which is called the *conjunction*† of p and q. It asserts *both* p and q.

† In most of the literature, the symbols $\wedge$ and $\vee$ rather than $\cdot$ and $+$ are used for conjunction and disjunction, respectively.

Similarly, we use $p + q$ to stand for the *disjunction*† of p and q, which is the statement

"EITHER 2 is an even integer OR π is rational (or both)".

It asserts *at least one* of p and q (and possibly both).

These notations naturally extend to any number of statements, $p, q, r, s, \cdots$; "$p \cdot q \cdot r \cdot s \cdot t$" asserts each of its components (i.e., all of them, then), while "$p + q + r + s + t$" asserts at least one of its components.

Two statements, whether compound or simple, are said to be equal, or *equivalent*, if they *mean* the same thing from a logical viewpoint—i.e., if either may be replaced by the other without affecting an argument. On this basis, we see that the distributive law of multiplication (conjunction) over addition (disjunction) holds in this algebra of statements:

$$p \cdot (q + r) = (p \cdot q) + (p \cdot r).$$

The left side asserts both p and $q + r$. This is to assert p and, in addition, at least one of q and r. Accordingly, this gives either p and q or p and r (or both), which is precisely what the right side asserts. This distributive law can easily be extended to the identity

$$p \cdot (q_1 + q_2 + \cdots + q_n) = p \cdot q_1 + pq_2 + \cdots + pq_n.$$

Now assertions are simply claims which may or may not really be true. We indicate that a statement p really is true by writing $p = 1$. If we know that p is false, we write $p = 0$.

Consider now an extended product $p \cdot q \cdot r \cdot s \cdots k$. This asserts every single one of its components. If all of them are true, that is, if $p = 1$, $q = 1$, $r = 1$, $\cdots$, $k = 1$, we have $p \cdot q \cdot r \cdot s \cdots k = 1$; if one of them (or more) is false we have $p \cdot q \cdot r \cdot s \cdots k = 0$. That is, $p \cdot q \cdot r \cdot s \cdots k = 0$ if and only if one of the components has a *truth-value* zero.

Now in ordinary arithmetic, we know that the product of a set of integers is 0 if and only if one of the integers is zero. Consequently,

† In most of the literature, the symbols $\wedge$ and $\vee$ rather than $\cdot$ and $+$ are used for conjunction and disjunction, respectively.

we have the result:

> *If each letter in* $p \cdot q \cdot r \cdot s \cdots k$ *is replaced by its truth-value, and the resulting product computed as in ordinary arithmetic, the number obtained is the truth-value of the compound statement* $p \cdot q \cdot r \cdot s \cdots k$.

We inquire now whether such an arithmetic rule applies also to an extended sum $p + q + r + \cdots + k$. This sum is true if and only if at least one of $p, q, r, \cdots, k$ is true. It is false, then, only when every single component is false; that is, we have $p + q + r + \cdots + k = 0$ if and only if $p = 0, q = 0, r = 0, \cdots, k = 0$. So far, then, the rule seems to work fine. If exactly one of $p, q, r, \cdots, k$ is true, the rule again works fine. But if more than one of the components really is true, the sum of the truth-values will amount to more than 1. And we have no meaning for such a thing as $p + q + \cdots + k = 5$. Accordingly, we make the following convention: the sum of any number of 1's is to be evaluated as just the number 1. On this basis, our rule may be applied to sums as well as products.

Because of the distributive law, any compound statement may be multiplied out to give a sum of products.

> *The truth-value of any statement is obtained by substituting for the components their individual truth-values and computing the resulting products and sums as in ordinary arithmetic, observing the single exception that the sum of any number of* 1*'s is to be taken as just* 1.

Although these considerations lead to more serious mathematics we shall content ourselves here with cracking an old chestnut which usually goes something like this:

Lucy, Minnie, Nancy, and Opey ran a race. Asked how they made out, they replied:

Lucy: "Nancy won; Minnie was second."
Minnie: "Nancy was second and Opey was third."
Nancy: "Opey was last; Lucy was second."

If each of the girls made one and only one true statement, who won the race?

We employ the following notation:

> L_1 represents the statement "Lucy was first";
> L_2 represents the statement "Lucy was second";
> L_3 represents the statement "Lucy was third";
> L_4 represents the statement "Lucy was fourth";
> M_1 represents the statement "Minnie was first";

and so on. In these terms, Lucy made the statements N_1 and M_2. Now because only one of these is true, it must be that "N_1 is true and M_2 false" or "N_1 is false and M_2 is true". In symbols, that is to say, $(N_1 \cdot \sim M_2) + (\sim N_1 \cdot M_2)$ (recalling that the symbol $\sim$ negates, or falsifies, a statement). Since each girl made one and only one true statement, this compound statement really is true. Hence we may write

$$(N_1 \cdot \sim M_2) + (\sim N_1 \cdot M_2) = 1.$$

In the same way, the replies of Minnie and Nancy give

$$(N_2 \cdot \sim O_3) + (\sim N_2 \cdot O_3) = 1$$

and

$$(O_4 \cdot \sim L_2) + (\sim O_4 \cdot L_2) = 1.$$

Now if we know that statements $x,\ y,\ z$ are all true, i.e., that $x = 1,\ y = 1,\ z = 1$, then we know that the compound statement $x \cdot y \cdot z$, which claims all of them, is true, i.e., $x \cdot y \cdot z = 1$. Accordingly, we have

$$\begin{aligned}[(N_1 \cdot \sim M_2) + (\sim N_1 \cdot M_2)] \cdot [(N_2 \cdot \sim O_3) + (\sim N_2 \cdot O_3)]& \\ \cdot [(O_4 \cdot \sim L_2) + (\sim O_4 \cdot L_2)] &= 1.\end{aligned}$$

By the distributive law we can multiply out to get

$$\begin{aligned}&N_1 \cdot \sim M_2 \cdot N_2 \cdot \sim O_3 \cdot O_4 \cdot \sim L_2 + N_1 \cdot \sim M_2 \cdot N_2 \cdot \sim O_3 \cdot \sim O_4 \cdot L_2 \\ &+ N_1 \cdot \sim M_2 \cdot \sim N_2 \cdot O_3 \cdot O_4 \cdot \sim L_2 + N_1 \cdot \sim M_2 \cdot \sim N_2 \cdot O_3 \cdot \sim O_4 \cdot L_2 \\ &+ \sim N_1 \cdot M_2 \cdot N_2 \cdot \sim O_3 \cdot O_4 \cdot \sim L_2 + \sim N_1 \cdot M_2 \cdot N_2 \cdot \sim O_3 \cdot \sim O_4 \cdot L_2 \\ &+ \sim N_1 \cdot M_2 \cdot \sim N_2 \cdot O_3 \cdot O_4 \cdot \sim L_2 + \sim N_1 \cdot M_2 \cdot \sim N_2 \cdot O_3 \cdot \sim O_4 \cdot L_2 = 1.\end{aligned}$$

This tells us that at least one of these products really is true. Could it be the first one? Clearly not, because it claims that Nancy is both first and second, impossible. The second product is rejected for the same impossibility. In fact, equally obvious contradictions force us to reject every one of these products except the fourth one. Substituting the truth-value 0 for each of the other seven rejected products, we obtain that the truth-value of

$$N_1 \cdot \sim M_2 \cdot \sim N_2 \cdot O_3 \cdot \sim O_4 \cdot L_2$$

is 1. Consequently, every component in it must be true. We see, then, that

Nancy was first,

Lucy was second,

Opey was third, and

Minnie was fourth (by elimination).

NOTE: Everywhere in mathematics one encounters statements of the kind "if ···, then ···". These are called "implications" and the implication "if p, then q" is often denoted simply by $p \rightarrow q$. The following observation is most useful in applying our algebraic technique to problems in logic:

The implication $p \rightarrow q$ *translates into the equation* $\sim p + q = 1$.

We easily verify this as follows. The statement $\sim p + q$ asserts at least one of "p is false", "q is true". Clearly this is so, because the former holds when p is false and the latter when p is true (recall $p \rightarrow q$).

For example, in dealing with Exercise 1 below we can use the fact that

"if Clark is the shipper, Brown is the driver"

by forming the equation $\sim C_s + B_d = 1$ to conclude that either Clark is not the shipper or Brown is the driver".

EXERCISES

1. Three men named Arnold, Brown and Clark hold the positions of shipper, driver and manager in a certain industrial firm.

   If Clark is the shipper, Brown is the driver.
   If Clark is the driver, Brown is the manager.
   If Brown is not the shipper, Arnold is the driver.
   If Arnold is the manager, Clark is the driver.

   What is each man's occupation?

2. Given:
   If A is not guilty, then B and C are both guilty.
   Either A is not guilty or B is guilty.
   Either B is not guilty or C is not guilty.

   Where does the guilt lie?

3. A says "B is a liar or C is a liar";
   B says "A is a liar";
   C says "A is a liar and B is a liar".

   Who is telling the truth?

   (*Hint*: interpret "A says" to mean "the truthfulness of A implies".)

4. What logical conclusion can be drawn from:

   A says B and C both tell the truth;
   B says A tells the truth;
   C says A and B are both liars?

5. Out of six boys, two were known to have been stealing apples. But who? Harry said "Charlie and George". James said "Donald and Tom". Donald said "Tom and Charlie". George said "Harry and Charlie". Charlie said "Donald and James". Tom couldn't be found. Four of the boys interrogated named one miscreant correctly, and one incorrectly. The fifth had lied outright. Who stole the apples?†

† Exercises 5, 6, 7 appear in an expository paper by J. T. Fletcher, Mathematical Gazette, 1952, pp. 183–188.

6. Three counters, A, B, C, are coloured red, white and blue, but not necessarily respectively. Of the following statements, one only is true:

   A is red; B is not red; C is not blue.

   What colour is each counter?†

7. Four members of my club—Messrs. Albert, Charles, Frederick, and Dick—have recently been knighted, so now their friends have had to learn their Christian names. These are a bit troublesome; for it transpires that the surname of each of the four knights is the Christian name of one of the others. Dick is not the Christian name of the member whose surname is Albert. There are three of the knights related as follows: the Christian name of the member whose surname is Frederick is the surname of the member whose Christian name is the surname of the member whose Christian name is Charles. What is Mr. Dick's first name?†

REFERENCE

D. Pedoe, *The Gentle Art of Mathematics*, (Reprint) Penguin, 1969, Baltimore.

† Exercises 5, 6, 7 appear in an expository paper by J. T. Fletcher, Mathematical Gazette, 1952, pp. 183–188.

ESSAY FIVE

The Farey Series

The proper fractions, in their lowest terms, may be written in a series of rows as illustrated in Table 5.1. Each row begins with 0/1 and ends with 1/1, its entries being written from left to right in order of increasing magnitude. The proper fractions on the *n*-th row, called the *n*-th Farey Series (or, more properly, the *n*-th Farey Sequence), are all those with denominator equal to or less than *n* (see Table 5.1).

$$F_1:\quad \frac{0}{1}\ \frac{1}{1}$$

$$F_2:\quad \frac{0}{1}\ \frac{1}{2}\ \frac{1}{1}$$

$$F_3:\quad \frac{0}{1}\ \frac{1}{3}\ \frac{1}{2}\ \frac{2}{3}\ \frac{1}{1}$$

$$F_4:\quad \frac{0}{1}\ \frac{1}{4}\ \frac{1}{3}\ \frac{1}{2}\ \frac{2}{3}\ \frac{3}{4}\ \frac{1}{1}$$

$$F_5:\quad \frac{0}{1}\ \frac{1}{5}\ \frac{1}{4}\ \frac{1}{3}\ \frac{2}{5}\ \frac{1}{2}\ \frac{3}{5}\ \frac{2}{3}\ \frac{3}{4}\ \frac{4}{5}\ \frac{1}{1}$$

$$F_6:\quad \frac{0}{1}\ \frac{1}{6}\ \frac{1}{5}\ \frac{1}{4}\ \frac{1}{3}\ \frac{2}{5}\ \frac{1}{2}\ \frac{3}{5}\ \frac{2}{3}\ \frac{3}{4}\ \frac{4}{5}\ \frac{5}{6}\ \frac{1}{1}$$

$$F_7:\quad \frac{0}{1}\ \frac{1}{7}\ \frac{1}{6}\ \frac{1}{5}\ \frac{1}{4}\ \frac{2}{7}\ \frac{1}{3}\ \frac{2}{5}\ \frac{3}{7}\ \frac{1}{2}\ \frac{4}{7}\ \frac{3}{5}\ \frac{2}{3}\ \frac{5}{7}\ \frac{3}{4}\ \frac{4}{5}\ \frac{5}{6}\ \frac{6}{7}\ \frac{1}{1}$$

....

etc.

Table 5.1

This table of numbers has (among others) the following two remarkable properties:

Property 1: For any pair of consecutive fractions a/b and c/d, the value of $bc - ad$ is 1.

For example, for the pair 1/2 and 2/3 in row 3,

$$2(2) - 1(3) = 4 - 3 = 1;$$

for the pair 4/7 and 3/5 in row 7,

$$7(3) - 4(5) = 21 - 20 = 1.$$

Property 2: For any three consecutive fractions a/b, c/d and e/f, the value of the middle one, c/d, is equal to the quotient $(a+e)/(b+f)$ of the sums of the numerators and denominators of the other two.

For example, for the triple 3/4, 4/5, 1/1 of row 5,

$$\frac{3+1}{4+1} = \frac{4}{5};$$

for the triple 5/7, 3/4, 4/5 of row 7,

$$\frac{5+4}{7+5} = \frac{9}{12} = \frac{3}{4}.$$

It is our purpose in this section to establish these properties from a geometric approach. We begin by showing that we need prove only the first property because the second follows from it.

From the consecutive trio a/b, c/d, e/f we obtain, by Property 1, that $bc - ad = 1$ and $de - cf = 1$. Therefore

$$bc - ad = de - cf,$$

and

$$bc + cf = de + ad.$$

This yields

$$c(b+f) = d(e+a),$$

from which we conclude that

$$\frac{c}{d} = \frac{a + e}{b + f}.$$

Thus the first property implies the second.

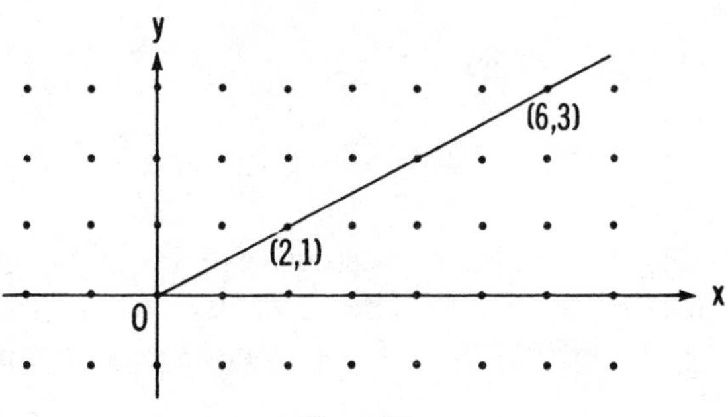

Figure 5.1

To prove Property 1, we use a set of perpendicular coordinate axes in the plane and single out for consideration those points whose coordinates are both integers. Such points are called *lattice points*. We distinguish two classes of lattice points: those which can be seen from the origin, and which we shall call *visible*, and those which are blocked from view by those in front of them, like trees in an orchard, and which we shall call *hidden*. [See Figure 5.1; for example, $(2, 1)$ is a visible point, $(6, 3)$ is a hidden point.] A little thought shows that the point (ka, kb) whose coordinates have a common factor $k > 1$ is hidden by the point (a, b). From this we see that *the coordinates of a visible point are relatively prime numbers*. We establish next the converse of this: *if* a *and* b *are relatively prime numbers, then the point* (a, b) *is visible.*

Since the visible and hidden points on the axes are easily sorted out, we consider non-zero integers a and b. Suppose that a and b are relatively prime, but that $P(a, b)$ is hidden by $Q(c, d)$. By similar triangles,

$$\frac{c}{d} = \frac{a}{b}$$

(see Figure 5.2). Now, since Q is closer to the origin than P, and since a and c are different (P is *not* on the y-axis), the magnitude of c is smaller than that of a. Likewise, d is smaller in magnitude than b. But a/b and c/d are equal, so a/b can be reduced to lower terms. That is to say, a and b are *not* relatively prime numbers. This contradiction establishes our result.

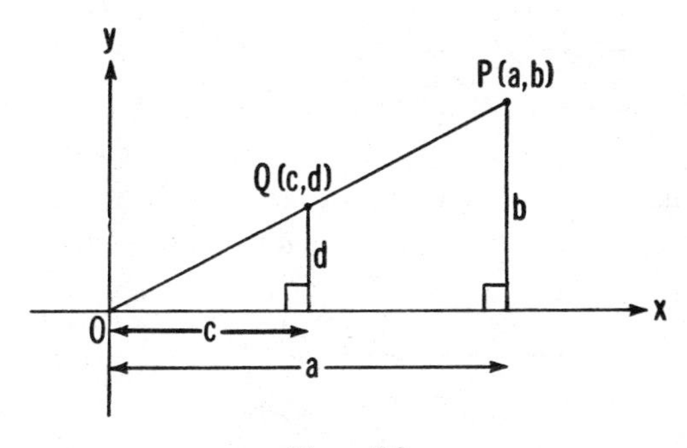

Figure 5.2

We need one other result from lattice geometry—Pick's Theorem. This is of interest in its own right and important in deriving many other results. We therefore interrupt our discussion to state and prove Pick's Theorem.

PICK'S THEOREM. *Let Π be a simple polygon† whose vertices lie on lattice points. Suppose there are q lattice points in the interior of Π and p lattice points on its boundary. Then*

$$\text{(1)} \qquad \text{Area of } \Pi = q + \frac{p}{2} - 1.$$

Proof: We show first that Pick's Theorem holds for a triangle whose vertices are lattice points but which has no other lattice points either in its interior or on its sides. We shall call such a triangle a ***primitive triangle***. For a primitive triangle, $p = 3$, $q = 0$, so formula (1) would yield the area $1/2$. Thus we must prove that the area of any primitive triangle is $1/2$.

† A simple polygon is one that does not cross itself.

We recall, from analytic geometry, that the area of a triangle with vertices (x_1, y_1), (x_2, y_2), (x_3, y_3) is given by the absolute value† of

$$(2)\quad \frac{1}{2}\begin{vmatrix} x_1 & y_1 & 1 \\ x_2 & y_2 & 1 \\ x_3 & y_3 & 1 \end{vmatrix} = \tfrac{1}{2}(x_1y_2 + x_2y_3 + x_3y_1 - x_1y_3 - x_2y_1 - x_3y_2).$$

Clearly, if the vertices are lattice points, all the x's and y's are integers, and so are their products, and hence the expression in brackets in the right member of (2) is an integer. This integer is zero only if the three points are collinear (in which case we have no triangle), so we conclude that this integer is at least 1. It follows that *the area of a primitive triangle is at least* 1/2.

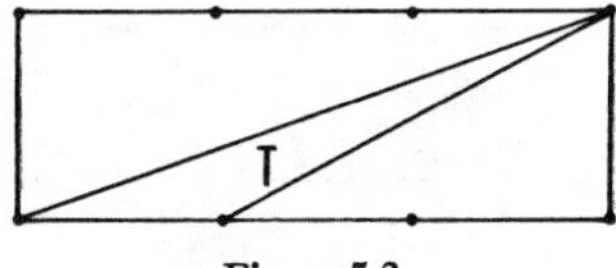

Figure 5.3

To show that it is exactly 1/2, enclose the primitive triangle T in a rectangle R as follows: One pair of bounding edges of the rectangle lie on the vertical lines through the left-most and right-most vertex, respectively, of T (see Figure 5.3). The other pair of edges lie on the horizontal lines through the uppermost and the lowest vertex, respectively, of T. The rectangle R so obtained has a number, say k, of lattice points in its interior, a number, say l, on its boundary of which four are at the vertices of R and $l - 4$ on other parts of the edges. It is possible to partition R into primitive triangles (see Appendix). Suppose our partition of R contains n primitive triangles, one of them being T. We claim that *any partition of R into primitive triangles contains the same number of primitive triangles.* We establish this assertion by a count of angles.

Each interior lattice point of R is a common vertex of some primitive triangles. Their vertex angles at such a point add up to 360° [see Figure 5.4(a)]. Since there are k interior lattice points, this accounts for $k \cdot 360$ degrees.

† (2) is positive if the vertices (x_1, y_1), (x_2, y_2), (x_3, y_3) appear in the counterclockwise sense and negative otherwise.

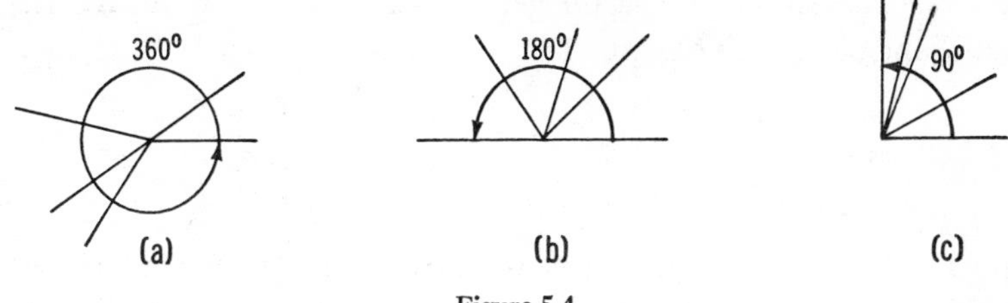

Figure 5.4

At each lattice point on an edge (but not at a corner) of R, the primitive triangles contribute 180° [see Figure 5.4(b)] and since R has $l - 4$ such lattice points, this accounts for $(l - 4)\cdot 180$ degrees.

At each corner of R, the primitive triangles contribute 90° [Figure 5.4(c)], so we have $4\cdot 90$ degrees from the four corners. Having thus accounted for all angles of the n primitive triangles in a given partition, we conclude that their sum is

$$k\cdot 360 + (l - 4)\cdot 180 + 4\cdot 90$$

degrees. On the other hand, the sum of the angles in each partition triangle is 180°, and since there are n of them, we have

$$180n = 360k + (l - 4)180 + 4\cdot 90,$$

whence

$$n = 2k + l - 2.$$

Thus n depends only on k and l, not on the particular way of partitioning R into primitive triangles.

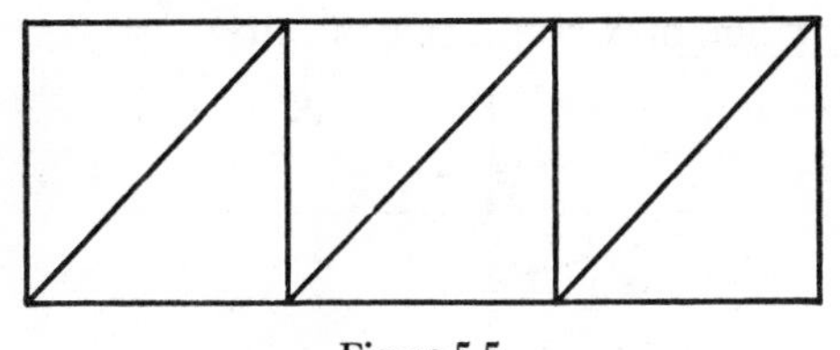
Figure 5.5

We now partition R into a set of congruent primitive triangles, each of area $1/2$, by first drawing all the lattice squares in R and then drawing a diagonal of each such square (see Figure 5.5). By what we have just shown, we again get $n = 2k + l - 2$ primitive triangles, and since each has area 1/2,

$$\text{Area } (R) = \frac{n}{2}.$$

Let $T_1, T_2, \cdots, T_n$ be the n primitive triangles in our original partition, where one of them was the given primitive triangle. Denote the area of T_i by (T_i). Then, by what we learned on page 28,

$$(3) \qquad (T_i) \geq \tfrac{1}{2} \qquad \text{for } i = 1, 2, \cdots, n;$$

on the other hand, since the T_i fill out R,

$$(4) \qquad \sum_{i=1}^{n} (T_i) = (T_1) + (T_2) + \cdots + (T_n) = \frac{n}{2}.$$

Equations (3) and (4) clearly imply that

$$(T_1) = (T_2) = \cdots = (T_n) = \tfrac{1}{2}.$$

In particular,

$$(T) = \tfrac{1}{2}.$$

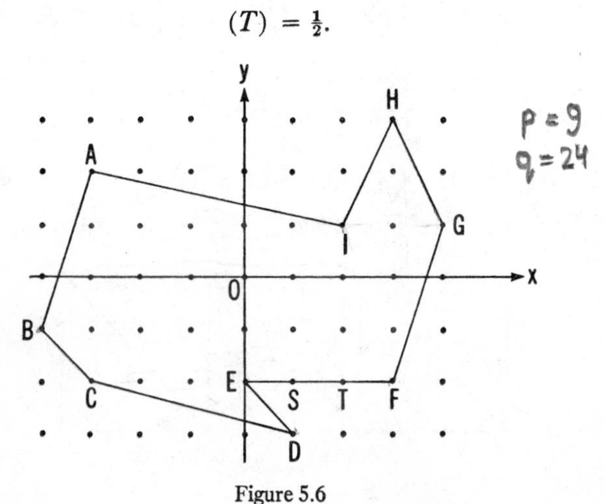

Figure 5.6

Now suppose we have a simple polygon with p lattice points on its boundary and q lattice points in its interior (see Figure 5.6). It is shown in the Appendix that any such polygon can be partitioned into primitive triangles. Again the number of primitive triangles in any such partition is always the same, as we shall now show by imitating the method we used earlier for a rectangle.

Suppose our polygon has v vertices, each, of course, a lattice point. Then it has $p - v$ lattice points on its sides (but not at their end-points). Now the primitive triangles in the partition contribute the following angle sums, in degrees:

$360q$ at the interior lattice points,

$180(p - v)$ at the boundary lattice points exclusive of those at the vertices,

$(v - 2)180$ at the vertex lattice points, since the sum of the interior angles of a v-gon is $(v - 2)180$.

If there are n primitive triangles in the partition, the sum of all their angles is $180n$, so

$$180n = 360q + 180(p - v) + (v - 2)180$$

and

$$n = 2q + p - v + v - 2 = 2q + p - 2.$$

Since each of these n triangles has area $1/2$, our polygon has area

$$\frac{n}{2} = q + \frac{p}{2} - 1$$

as asserted in Pick's Theorem.

We now return to the proof of Property 1 and associate the fraction a/b of a Farey Series with the lattice point (a, b). Because a/b is in its lowest terms (i.e., a and b are relatively prime), the point (a, b) is a visible point. The visible points corresponding to any particular Farey Sequence (which we denote by F_n) all occur in a certain region of the plane. For definiteness, let us consider F_4:

$$\frac{0}{1}\ \frac{1}{4}\ \frac{1}{3}\ \frac{1}{2}\ \frac{2}{3}\ \frac{3}{4}\ \frac{1}{1}.$$

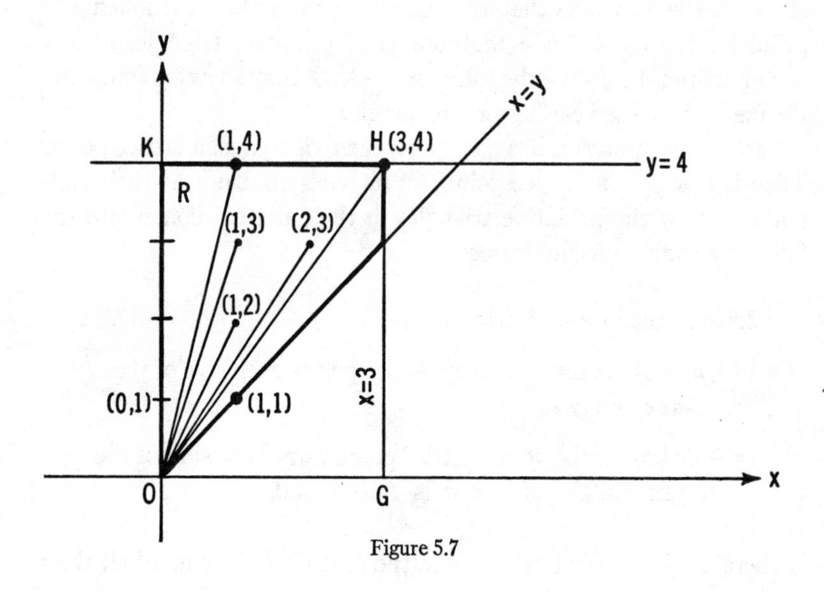

Figure 5.7

The corresponding lattice points are plotted in Figure 5.7.

Representing the fractions of F_4 by a/b, we have $b \leq 4$, and, since the fractions are proper (except 1/1), a can be at most 3. Consequently, the points (a, b) lie inside or on the rectangle with vertices

$$O(0,\ 0),\quad G(3,\ 0),\quad H(3,\ 4),\quad K(0,\ 4).$$

Because the fractions are proper (except 1/1), we have $a \leq b$, which tells us that these visible points lie on or above the line joining $(0,\ 0)$ and $(1,\ 1)$, i.e., the line $x = y$. As a result, the visible points of F_4 are confined to the interior and boundary of the quadrilateral R with sides (i) the y-axis, (ii) the line $x = y$, (iii) $x = 3$ and (iv) $y = 4$.

Next we show that the *only* visible points in the region R are those associated with F_4. Suppose $T(u,\ v)$ is a visible point in R. If T is on the y-axis, then it must be the point $(0,\ 1)$, corresponding to the fraction 0/1, which begins F_4; if T is on the line $x = y$, T must be $(1,\ 1)$, corresponding to the fraction 1/1 of F_4. If T is elsewhere in R, then

$u < v$ because T is above the line $x = y$;

$u \leq 4$ because T is on or below the line $y = 4$;

u and v are relatively prime, because T is visible.

As a result, the fraction u/v is a proper fraction with denominator equal to or less than 4, and it is in its lowest terms. By definition, then, it occurs in F_4.

We note that our argument obviously generalizes from F_4 to F_n.

Drawing the lines of sight from the origin to the visible points of F_n in R, we produce a "fan" stemming from the origin. (See Figure 5.7.) The slope of the line of sight to the point (a, b) is b/a, the reciprocal of the fraction of F_n under consideration. Because the fractions in F_n are written in order of increasing magnitude, the slopes of the corresponding lines of sight will occur in order of decreasing magnitude. Thus, consecutive fractions in the series give consecutive lines in the fan, and conversely.

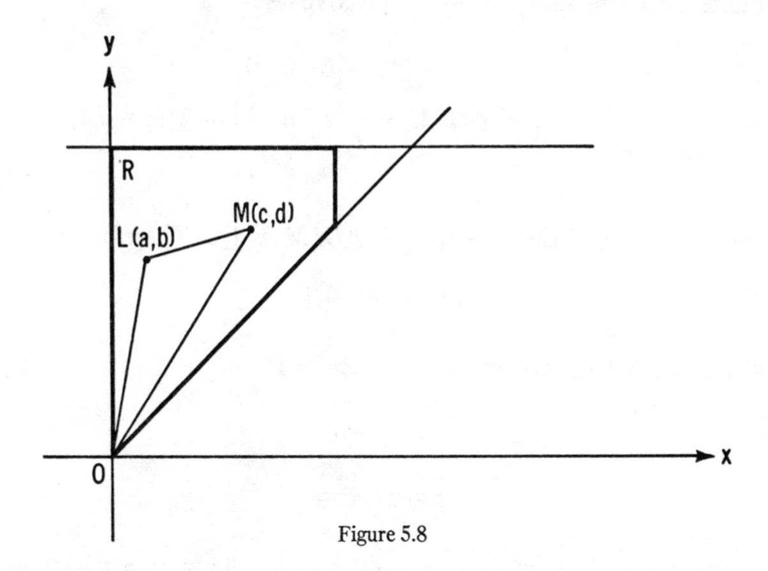

Figure 5.8

Now let a/b and c/d represent any pair of consecutive fractions in the series F_n. Let $L = (a, b)$ and $M = (c, d)$ be the corresponding lattice points (Figure 5.8). Because the fractions are consecutive in F_n, the lines OL and OM are *consecutive* in the fan of F_n. Therefore there is no visible point of F_n *inside* triangle OLM or on the side LM.

Clearly, no visible point can occur on OL between O and L because L, itself, is visible. Similarly for OM. Thus there is no visible point either in or on $\triangle OLM$ except the vertices. And, if there is no visible point, then there is no lattice point at all; for any hidden point would be hidden by a visible point between it and O. Consequently, $\triangle OLM$ is a primitive triangle, and so its area is 1/2. By Pick's formula with $p = 3, q = 0$, we get

$$\tfrac{3}{2} + 0 - 1 = \tfrac{1}{2}$$

for the area of $\triangle OLM$.

Finally, we compute the area of $\triangle OLM$ by the usual method of analytic geometry [see (2) on p. 28]. Since a/b precedes c/d in F_n, we have $(a/b) < (c/d)$, or $(b/a) > (d/c)$, or

$$\text{slope of } OL > \text{slope of } OM.$$

Consequently, the circuit O, M, L, O traces the triangle in the counterclockwise sense, ensuring a positive result for

$$\text{Area } \triangle OLM = \frac{1}{2}\begin{vmatrix} 0 & 0 & 0 \\ c & d & 1 \\ a & b & 1 \end{vmatrix} = \tfrac{1}{2}(bc - ad).$$

But we just found that the area of $\triangle OLM$ is 1/2. Therefore

$$\tfrac{1}{2}(bc - ad) = \tfrac{1}{2},$$

whence we deduce Property 1, $bc - ad = 1$.

REFERENCES

H. S. M. Coxeter, *Introduction to Geometry*, (2nd ed.), Wiley, 1969, New York.

H. Rademacher, *Lectures in Elementary Number Theory*, Blaisdell, 1964, New York.

Enrichment Mathematics for High Schools, NCTM Yearbook 28, 1963, Washington, D.C.

Appendix

THEOREM. *Every simple polygon can be decomposed into triangles by means of non-intersecting diagonals lying in the interior of the polygon.*

Proof:† We shall prove this by mathematical induction on the number of vertices v of the polygon. If $v = 4$, the theorem is true (see Figure 5.9).

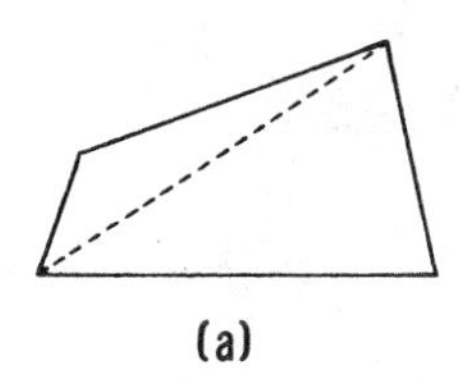

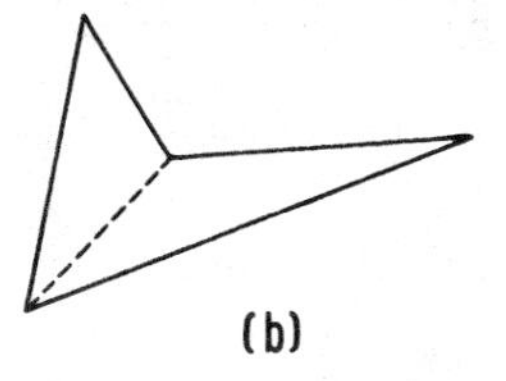

Figure 5.9

Let $v > 4$, and suppose the theorem is true for polygons with fewer than v vertices. To prove that the theorem is true for any polygon Π with v vertices, it would suffice to show that *there is an interior diagonal which divides* Π *into two subpolygons*, Π_1 *and* Π_2. For since any subpolygon of Π has at most $v - 1$ vertices, Π_1 and Π_2 satisfy our induction hypothesis.

To show that Π has at least one interior diagonal, we draw a line l which has no points in common with Π. This can always be done since Π occupies some bounded region in the plane. We now move l parallel to itself toward Π until Π and l have a point in common. In this position l necessarily contains a vertex, say A, of Π, and the interior angle of Π at A is less than a straight angle. Let B and C be the vertices adjacent to A. Then precisely one of the following must occur:

1) BC is an interior diagonal of Π [see Figure 5.10(a)];

2) there is at least one other vertex of Π on BC, but no vertex inside $\triangle ABC$ [see Figure 5.10(b)];

3) there is at least one vertex of Π inside $\triangle ABC$ [see Figure 5.10(c)].

† This proof follows the one given by Konrad Knopp in *Theory of Functions*, I, Dover, New York, 1945, pp. 17–18.

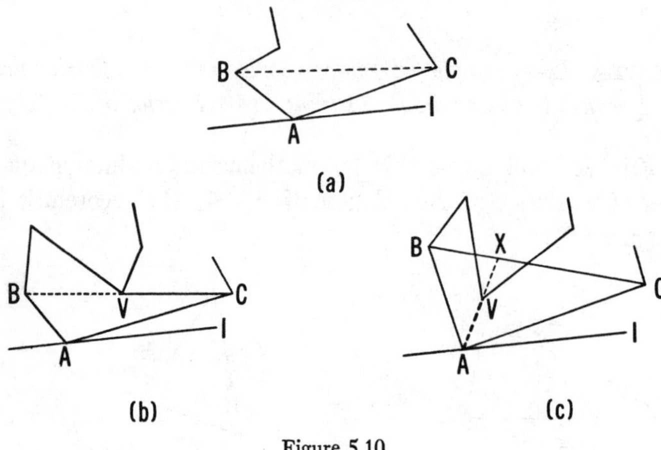

Figure 5.10

In case 1), we have the desired interior diagonal. In case 2), the segment BV from B to the first vertex (say V) encountered on the path to C is an interior diagonal. In case 3) let a point X move from B along BC toward C, and consider the segments XA which sweep out $\triangle ABC$. The first such segment AX which encounters a vertex (it may go through several vertices) is divided into two subsegments: AV, from A to the first vertex V on AX, and VX, the remaining interval. AV is then an interior diagonal of Π.

This completes the proof that any simple polygon Π may be decomposed into triangles whose vertices are vertices of Π.

Next we show that a simple polygon Π whose vertices are lattice points can be decomposed into primitive triangles. In view of what we have just proved, it suffices to demonstrate that *every triangle whose vertices are lattice points can be decomposed into primitive triangles.*

Proof: Consider a triangle whose vertices A, B, C are lattice points. If $\triangle ABC$ has no other lattice point either on its sides or in its interior, $\triangle ABC$ is primitive, and there is nothing to prove.

(a) Suppose $\triangle ABC$ has other lattice points on its sides, but none in its interior. Let $C_1, C_2, \cdots, C_k$ be lattice points on side AB (see Figure 5.11), and connect each with the opposite vertex C.

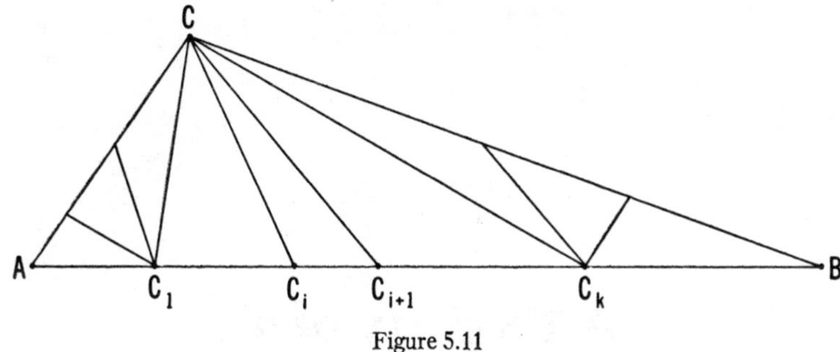

Figure 5.11

Then all triangles CC_iC_{i+1} $(i = 1, \cdots, k-1)$ are primitive, while triangles ACC_1 and BCC_k have lattice points at most on sides AC and BC respectively. We connect those on AC with C_1, those on BC with C_k and so arrive at a decomposition of $\triangle ABC$ into primitive triangles.

(b) Suppose $\triangle ABC$ has interior lattice points. We first decompose it into subtriangles having no interior lattice points (by induction on the number of interior lattice points); these subtriangles then satisfy the hypothesis of (a) above and are further decomposed into primitive triangles.

If $\triangle ABC$ has one interior lattice point V, we connect V to A, to B, and to C, obtaining three triangles free of interior lattice points. Suppose that a triangle with fewer than n interior lattice points can be so decomposed. In a triangle with n interior lattice points, pick any one of the n interior lattice points of $\triangle ABC$ and connect it to A, B and C. Each of the resulting triangles AVB, BVC, CVA has fewer than n interior lattice points and can therefore be decomposed into triangles free of interior lattice points. The method of part (a) will decompose these into primitive triangles, thus achieving our goal.

ESSAY SIX

A Property of a^n

If you were asked to name a power of 2 that begins with a 3, no doubt you would quickly reply "32" ($= 2^5$). Again, the request for a power of 2 that begins with 12 would soon bring "128" ($= 2^7$). But if you were asked for a power of 2 that begins with 11223344556677, you would likely remain silent. Indeed, you may well wonder whether such a power of 2 exists. The remarkable theorem that we prove in this essay settles this question; it asserts that *there exist powers of* 2 *beginning with any given sequence of digits.* As a matter of fact, the theorem makes the same claim for 3, 4, and any positive integer a which is not a power of 10 (i.e., $a \neq 1, 10, 100, 1000$, etc.). We prove the theorem for powers of 2; the general case is established in the same way.

Let $S = abc \cdots k$ be *any* sequence of digits. We need to show that, for some n,

$$2^n = abc \cdots k \cdots,$$

where there may be digits beyond the last digit of S.

What we want to consider first is the set of numbers which begin with the digits S. To fix the ideas, let us begin with a concrete example, say $S = 5$. (Here the required power is $2^9 = 512$.) If 2^n is to begin with a 5, then it must occur in one of the intervals

(i) $5 \leq 2^n < 6$, (i.e., $2^n = 5$)
(ii) $50 \leq 2^n < 60$,
(iii) $500 \leq 2^n < 600$,
(iv) $5000 \leq 2^n < 6000$,
(v) $50000 \leq 2^n < 60000$, etc.

Every integer in these intervals begins with a 5, and no others do. We may express the intervals as

$$\begin{array}{ll} \text{(i)} & 5\cdot 10^0 \leq 2^n < (5+1)\cdot 10^0, \\ \text{(ii)} & 5\cdot 10 \leq 2^n < (5+1)\cdot 10, \\ \text{(iii)} & 5\cdot 10^2 \leq 2^n < (5+1)\cdot 10^2, \\ \text{(iv)} & 5\cdot 10^3 \leq 2^n < (5+1)\cdot 10^3, \quad \text{etc.} \end{array}$$

In general, we have

$$\begin{array}{llll} & \text{(i)} & S \leq 2^n < (S+1) & (\text{i.e., } 2^n = S), \\ & \text{(ii)} & S\cdot 10 \leq 2^n < (S+1)\cdot 10, & \\ & \text{(iii)} & S\cdot 10^2 \leq 2^n < (S+1)\cdot 10^2, & \\ & \text{(iv)} & S\cdot 10^3 \leq 2^n < (S+1)\cdot 10^3, & \\ & \cdots\cdots & \cdots\cdots\cdots\cdots & \\ \text{(A)} & (t+1) & S\cdot 10^t \leq 2^n < (S+1)\cdot 10^t. & \end{array}$$

[$t = 0$ gives line (i), the obvious solution if S happens to be a power of 2, itself.] We must show that, given any S, there exist some t and n satisfying inequalities (A). We are not concerned with finding actual values for such t and n. We need only demonstrate that they exist.

Taking logarithms (to the base 10) throughout inequalities (A), we get

$$\text{(B)} \qquad t + \log S \leq n \log 2 < t + \log (S+1).$$

Clearly, any t and n which satisfy (B) also satisfy (A). We prove the theorem by showing that for any S there exist non-negative integers t and n satisfying (B).

First we mark the two numbers

$$m = \log S \qquad \text{and} \qquad l = \log (S+1)$$

on a number line to obtain the interval $[m, l)$† (see Figure 6.1).

† The interval $[a, b)$ consists of all points x such that $a \leq x < b$. We are interested in intervals containing the left but not the right endpoint because relation (B) asserts that $n \log 2$ is in such a "half-open" interval.

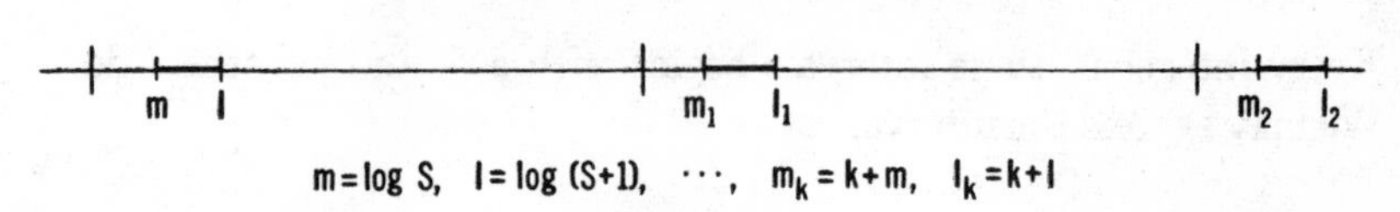

Figure 6.1

Since $S \geq 1$, the length of this interval is

$$
\begin{aligned}
l - m &= \log (S + 1) - \log S \\
&= \log \frac{S+1}{S} = \log\left(1 + \frac{1}{S}\right) < \log 2 < 1.
\end{aligned} \tag{1}
$$

Next, this interval is translated one unit to the right again and again, yielding the sequence of intervals

$$
\begin{aligned}
[m,\ l) &= [\log S,\ \log (S + 1)), \\
[m_1,\ l_1) &= [1 + \log S,\ 1 + \log (S + 1)), \\
[m_2,\ l_2) &= [2 + \log S,\ 2 + \log (S + 1)), \\
&\cdots\cdots\cdots\cdots\cdots\cdots\cdots\cdots \\
[m_i,\ l_i) &= [i + \log S,\ i + \log (S + 1)), \quad \text{etc.}
\end{aligned}
$$

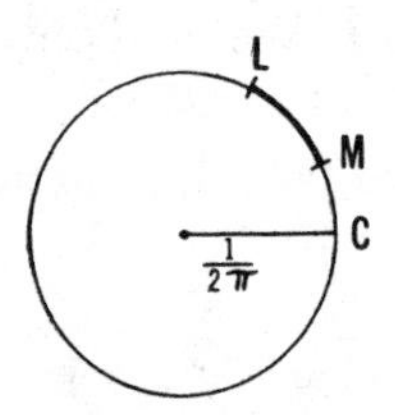

Figure 6.2

Now a circle of unit circumference (i.e., of radius $1/2\pi$) is constructed (see Figure 6.2), and the number line is wrapped around this circle in the counterclockwise direction so that its zero point 0 is mapped onto the point C of the circle. Of course, the positive real axis goes around the circle infinitely often. If two numbers differ by an integer, they are mapped on the same point of the circle, and conversely. Thus m, m_1, m_2, $\cdots$, as defined above, all map into M; and l, l_1, l_2, $\cdots$ all map into L.

Now, for any positive real number k, let $[k]$ denote its integer part; that is,

$$[k] = \text{the largest integer not exceeding } k.$$

For example,

$$[53.6] = 53, \qquad [14.3] = 14, \qquad [14] = 14.$$

Note that the length of a circular arc CK (measured counterclockwise from C to K) has the property

$$\text{length } \widehat{CK} = k - [k], \tag{2}$$

where points C, K on the circle correspond to the numbers 0, k. Observe also that

$$[\log (S+1)] = [\log S] \qquad \text{if } S+1 \text{ is not a power of 10;} \tag{3}$$

$$[\log (S+1)] = [\log S] + 1 \quad \text{if } S+1 \text{ is a power of 10.} \tag{3'}$$

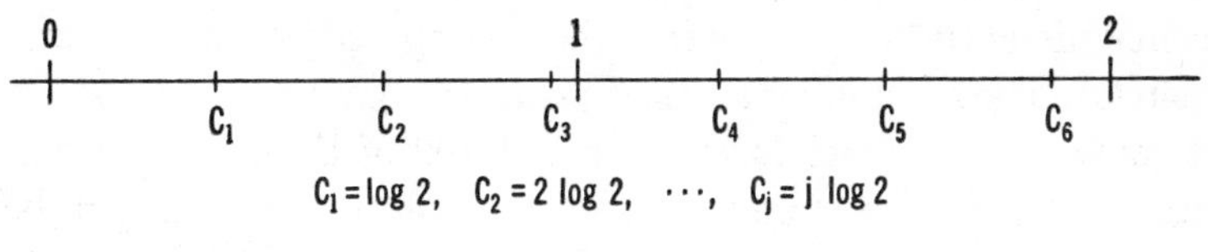

(a)

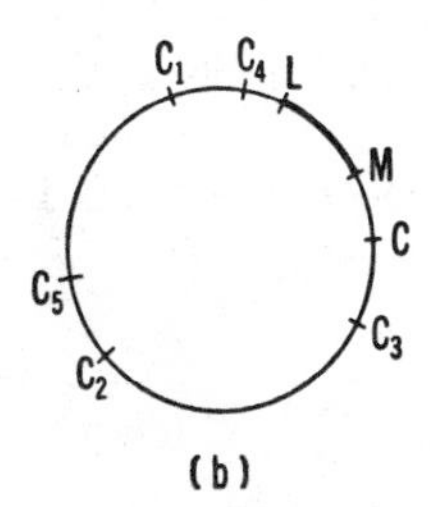

(b)

Figure 6.3

Next, consider the numbers

$$\log 2,\ 2 \log 2,\ 3 \log 2,\ \cdots,\ n \log 2,\ \cdots,$$

and denote their images on the circle by

$$C_1,\ C_2,\ C_3,\ \cdots,\ C_n,\ \cdots.$$

(See Figure 6.3.) They go around the circle in steps of arc length log 2. No two of them coincide; for, if C_i and C_j $(j > i)$ were mapped into the same point, the difference

$$j \log 2 - i \log 2 = (j - i) \log 2 = w$$

would be a positive integer, and this would imply that

$$\log 2 = \frac{w}{j - i}, \quad \text{a rational number.}$$

But log 2 is irrational†—a contradiction. Hence all points C_n are distinct.

Now, there are infinitely many of these points C_n. Consequently, there must be a pair of them whose distance apart is less than any preassigned number. If, for example, there were only 100 points C_n, it is conceivable that no two need be closer together than 1/100 of the circumference (in the case that they are spaced "evenly" around the circle). The addition of a 101-th point, though, would necessitate that some pair be closer together than 1/100 of the circumference. Similarly, if there are more than a million points, some pair will be closer together than one-millionth of the circumference. No matter how short the arc ML may be, sooner or later enough of the points C_n will be wrapped around the circle so that some pair of them will be separated by less than the length of $\widehat{ML}$. Let C_p and C_{p+q} (where $q > 0$) represent such a pair, i.e.,

$$\text{(4)} \qquad \text{length } \widehat{C_pC_{p+q}} < \text{length } \widehat{ML}.$$

(See Figure 6.4.)

† If log 2 were rational, say $\log 2 = u/v$, where u and v are integers, then we would have $2 = 10^{u/v}$, that is, $2^v = 10^u$. But this is impossible because the right side is divisible by 5 while the left side is not.

This is the only place that the nature of the base a of a^n enters into the argument; for the proof to work, a need only have an irrational common logarithm. By arguing in a manner similar to the case $a = 2$, above, any a which is not a power of 10 can be shown to satisfy this requirement. (The factors 2 and 5 are never equally distributed on each side of $a^v = 10^u$ unless a is a power of 10.)

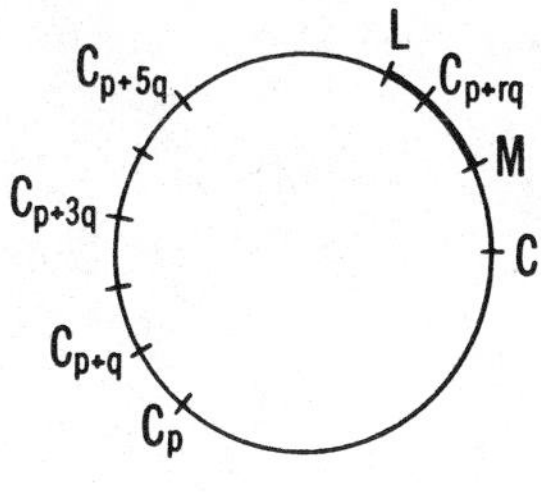

Figure 6.4

We do not know if the short arc from C_p to C_{p+q} is clockwise or counterclockwise, and it will not matter. We look at the points

$$C_p,\ C_{p+q},\ C_{p+2q},\ \cdots,\ C_{p+rq},\ \cdots$$

corresponding to the numbers

$$p \log 2,\ (p+q) \log 2,\ (p+2q) \log 2,\ \cdots,\ (p+rq) \log 2,\ \cdots$$

(an arithmetic progression with difference $q \log 2$) and observe that

$$\text{length } \widehat{C_pC_{p+q}} = \text{length } \widehat{C_{p+q}C_{p+2q}} = \cdots = \text{length } \widehat{C_{p+(r-1)q}C_{p+rq}} = \cdots;$$

in other words, these points march around the circle (either in clockwise or counterclockwise direction) in steps of arc length less than length $\widehat{ML}$ [see (4)], and therefore, each time they go around the circle, at least one of them must land in $\widehat{ML}$. We may conclude that, *given any number R no matter how large, there is an integer $r \geq R$ such that C_{p+rq} lies in $\widehat{ML}$.* (See Figure 6.4.)

Choose r so large that

$$(5) \qquad (p+rq) \log 2 \geq \log S.$$

In terms of arc length, we may write

$$(6) \qquad \text{length } \widehat{CM} \leq \text{length } \widehat{CC_{p+rq}} < \text{length } \widehat{CL},$$

and using (2), we rewrite this as

$$m - [m] \leq (p+rq) \log 2 - [(p+rq) \log 2] < l - [l],$$

that is,

$$\text{(7)}\qquad \begin{aligned}\log S - [\log S] \le (p + rq) \log 2 &- [(p + rq) \log 2] \\ &< \log (S + 1) - [\log (S + 1)].\end{aligned}$$

We add the integer

$$u = [(p + rq) \log 2]$$

to all members of (7) and obtain

$$\text{(8)}\qquad \begin{aligned}u + \log S - [\log S] &\le (p + rq) \log 2 \\ &< u + \log (S + 1) - [\log (S + 1)].\end{aligned}$$

It follows from (5) that the integer

$$t = u - [\log S]$$

is non-negative; if $S + 1$ is not a power of 10, we use (3) to deduce that $t = u - [\log (S + 1)]$. So in this case, (8) yields the desired relation

$$\text{(9)}\qquad t + \log S \le (p + rq) \log 2 < t + \log (S + 1).$$

In the rare case that $S + 1$ is a power of 10, we observe that the middle member of (7) denotes the fractional part of $(p + rq) \log 2$ and hence is less than 1, so

$$\text{(7}'\text{)}\qquad \log S - [\log S] \le (p + rq) \log 2 - u < 1.$$

On the other hand, in this case, (3′) yields

$$[\log (S + 1)] - [\log S] = 1,$$

and since now $\log (S + 1) = [\log (S + 1)]$, (7′) may be written as

$$\log S - [\log S] \le (p + rq) \log 2 - u < \log (S + 1) - [\log S].$$

Again we add u to all members, set $t = u - [\log S]$ and obtain the result (9) also in this exceptional case.

We conclude that, for any given S, there exist integers t and $n = p + rq$ such that inequalities (B) are satisfied. Thus the power 2^n begins with the digits of S.

The existence theorem we have just proved assures us that the search for a power of 2 beginning with any given digits is never futile; but unfortunately, the proof does not give us a recipe for constructing the desired power. It is worth noting that the smallest power of 2 beginning with 7 is 2^{46}, and the smallest power of 2 beginning with 9 is 2^{53}. Powers of 2 beginning with any other digit are very small in comparison:

Digit	1	2	3	4	5	6	8
Power	2^0, 2^4	2^8	2^5	2^2	2^9	2^6	2^3.

EXERCISE

Let a be any positive real number and n any positive integer; show that one of the numbers

$$a, \ 2a, \ 3a, \ \cdots, \ (n-1)a$$

differs from an integer by at most $1/n$.†

REFERENCE

A. M. Yaglom and I. M. Yaglom, *Challenging Mathematical Problems With Elementary Solutions*, Holden–Day, Inc., 1964, San Francisco.

† This problem and its solution appear in the Eötvös Competition of 1928 (see *Hungarian Problem Book II*, (NML vol. 12), Random House, 1963, New York).

ESSAY SEVEN

Squaring the Square

In 1936, four students at Trinity College, Cambridge—Brooks, Smith, Stone, and Tutte—considered the problem of cutting up a rectangle into squares of unequal size (no two alike). It was known, at that time, that a rectangle 32 by 33 could be "squared" as shown in Figure 7.1.

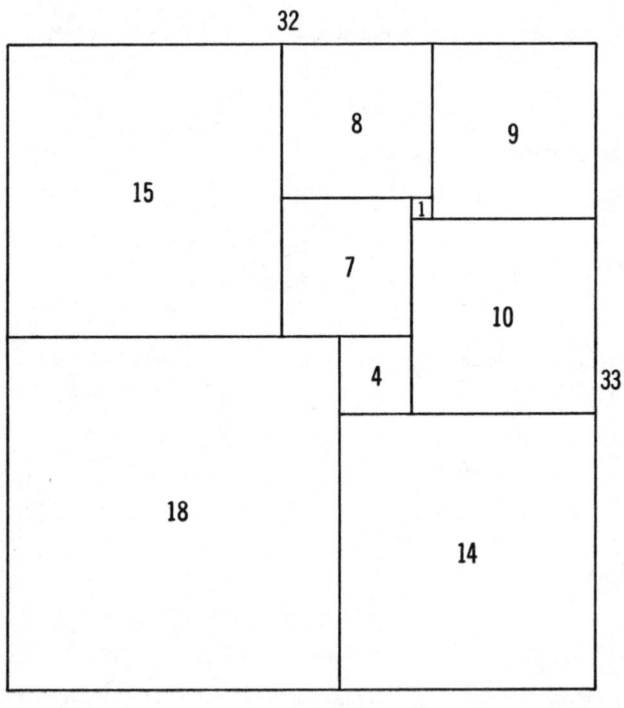

Figure 7.1

Stone became particularly interested in trying to prove that it was impossible to cut a given *square* into unequal squares. While not able to do this, he did discover a squaring of another rectangle (see Figure 7.2).

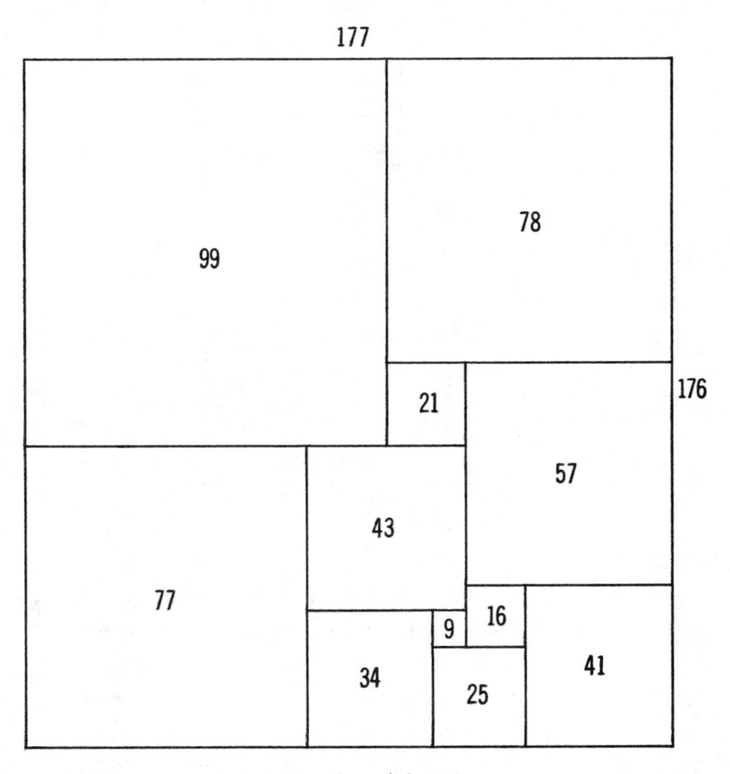

Figure 7.2

One way of attacking the problem of finding rectangles that can be squared is to make a sketch of a proposed partition into squares, labelling the edge-length of each square, writing down all the relations that the edge-lengths must satisfy in order to fit into the rectangle, and solving the system of equations thus obtained.

We shall carry this out for some examples. Instead of carrying as many unknowns as there are squares in the subdivision, we shall try to label neighbouring squares so that they fit together in the sketch; thus we shall have fewer unknowns to eliminate later.

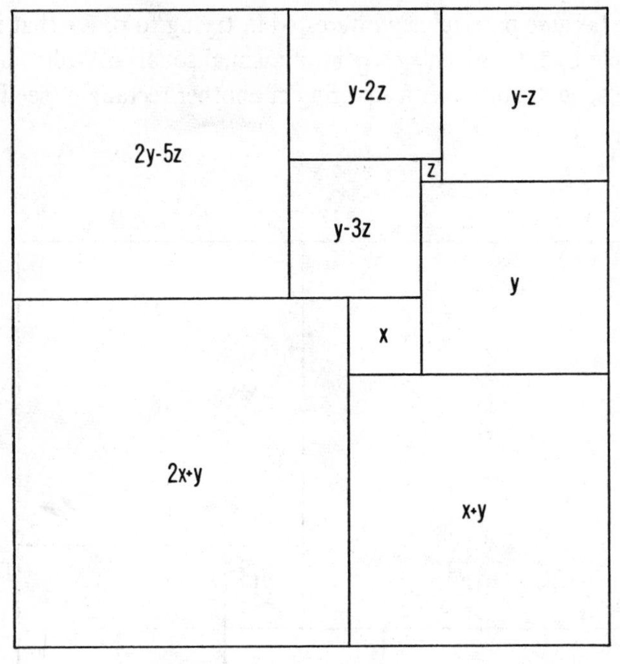

Figure 7.3

Example 1.

In Figure 7.3, we label the neighbouring squares x, y and z, as shown. Then it is easy to label the remaining squares in the order

$$x + y, \quad 2x + y, \quad y - z, \quad y - 2z, \quad y - 3z, \quad 2y - 5z.$$

Next, we obtain relations between our unknowns; for example, we can equate the lengths of opposite sides of the containing rectangle. The horizontal sides yield

$$2x + y + x + y = 2y - 5z + y - 2z + y - z,$$

that is,

$$(1) \qquad 3x - 2y + 8z = 0;$$

and the vertical sides yield

$$2y - 5z + 2x + y = y - z + y + x + y,$$

that is,

(2) $$x - 4z = 0.$$

So

$$x = 4z \qquad \text{and} \qquad y = 10z.$$

We observe that if we set $z = 1$, we obtain the tiling shown in Figure 7.1; if we set z equal to any other positive quantity, we obtain the same configuration, blown up (or shrunk) by the factor z. If we prescribe the length of the horizontal edge, say 64″, then z would be 2″ and the figure would be completely determined.

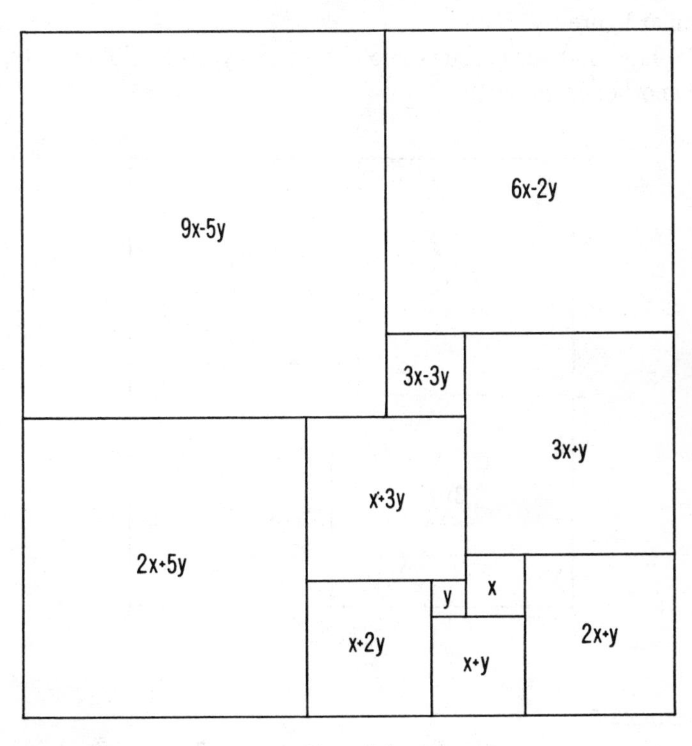

Figure 7.4

Example 2.

In Figure 7.4, two unknowns, x and y, suffice to express the edge-lengths of all compartments.

Equating the lengths of the horizontal sides of the containing rectangle, we obtain

$$9x - 5y + 6x - 2y = 2x + 5y + x + 2y + x + y + 2x + y,$$

that is, $9x - 16y = 0$, whence

$$\frac{x}{y} = \frac{16}{9}. \tag{3}$$

The values $x = 16$, $y = 9$ yield Stone's tiling of the 177×176 rectangle (see Figure 7.2), and other values satisfying (3) would give a similar figure.

To show that such schemes do not always yield feasible results, we give another example.

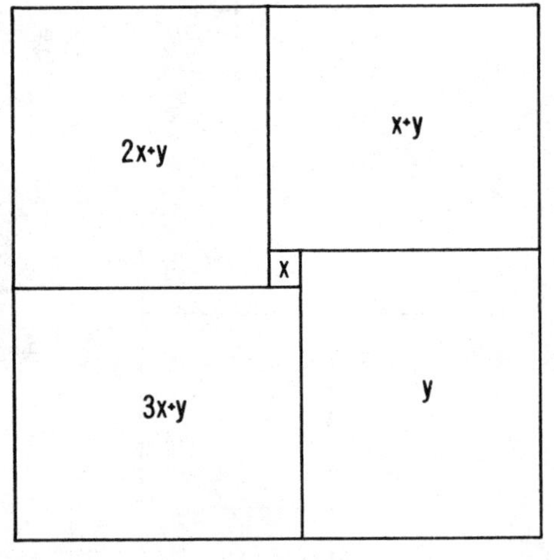

Figure 7.5

Example 3.

Starting with x and y as shown in Figure 7.5, we can label the other squares

$$x + y, \qquad 2x + y, \qquad 3x + y;$$

and equating lengths of the vertical borders, we get

$$\begin{aligned} 2x + y + 3x + y &= x + y + y \\ 5x + 2y &= x + 2y \\ 4x &= 0, \qquad \text{so } x = 0. \end{aligned}$$

Our middle square turns out to be a point and the remaining four squares are equal. We have merely quartered the given rectangle, which must have been a square!

Our examples indicate that the system of linear equations we get from an arbitrary sketch of a partition seems to have a unique solution (except for a scaling factor), although the solution is not necessarily geometrically feasible. It could happen, for example, that some of the edge-lengths turn out to be negative, and that would not make sense in the context of our tiling problem.

From our previous experience with systems of linear equations we know that there may be *too many* equations (in which case there is no solution) or there may be *too few* (in which case there are infinitely many solutions) or the number of equations may be *just right*, that is, they are satisfied by exactly one set of numbers. In all our examples, the number of equations obtained by prescribing the length of one pair of edges of the containing rectangle, but not the length of the other, was just right; i.e., each system had a unique solution. Was this just a happy accident?

We shall show, by appealing to electrical network theory, that a system of linear equations obtained from a sketched partition always has a unique solution.

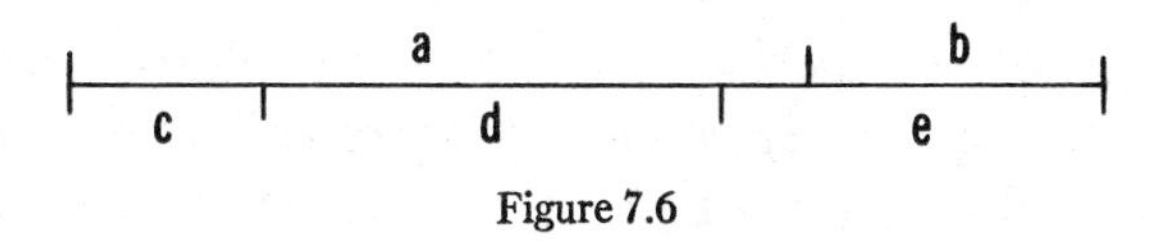

Figure 7.6

To make things more precise, consider a given subdivision of a rectangle into squares. With respect to each horizontal segment of the partition (see Figure 7.6) we have a relation

$$a + b = c + d + e$$

telling us that the sum of the lengths of the edges of squares bordering the segment on one side is equal to the sum of the lengths of the edges of squares bordering the segment on the other side. We shall call all such relations *horizontal compatibility relations*. Similarly, the analogous relations obtained from vertical segments will be called *vertical compatibility relations*. Clearly, the set of all compatibility relations forms a system S of linear equations.

THEOREM. *If one of the two dimensions of a partitioned rectangle is prescribed, the system* S *of compatibility relations always has a unique solution (though not necessarily geometrically meaningful).*

Imagine that the rectangle is a plate of uniform thickness made of some conducting metal. Suppose all points of the upper edge of the rectangle were given the same electric potential V, and all points on the lower edge had a lower constant potential V'. (This could be accomplished by coating these edges with some perfect conductor.) Because of the prescribed potential difference $V - V'$, there will be a steady flow of current through the rectangle, in the vertical direction. The rate at which electrons cross over a horizontal interval is proportional to the length of that interval. Thus, if I is the current crossing over a horizontal interval of unit length, then the current flowing across an interval of length l is lI. The *resistance* of such a rectangle to electric current is directly proportional to its vertical height L (to the distance the current must traverse) and inversely proportional to its horizontal width W (the distance along which the current may enter); i.e., resistance $= (RL/W)$. Thus, if the rectangle happens to be a square $(L = W)$, then its resistance is R and does not depend on the size of the square.

Now suppose that such a current-carrying rectangle has been partitioned into squares. Since the flow is vertical (there is no flow in the horizontal direction), we may cut through vertical segments without interfering with the current. Now regard the cut-up plate as a *network* where the constituent squares are identified *with conducting wires* and where horizontal segments are identified with points, or *vertices* at which the conducting wires meet. (See Figure 7.7, where the vertices $A, B, \cdots$ in the network correspond to horizontal segments at levels $a, b, \cdots$ of the rectangle, wires $AB, AC, \cdots$ correspond to squares extending from levels a to b, a to c, $\cdots$.) The magnitude of the current flowing through each wire is proportional to the length of the side of the square represented by that wire.

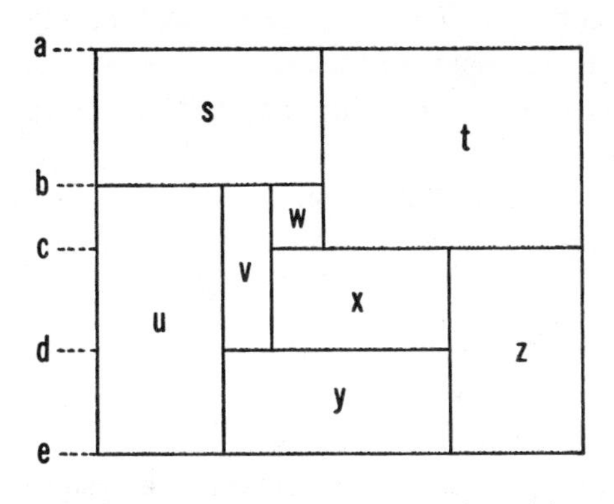

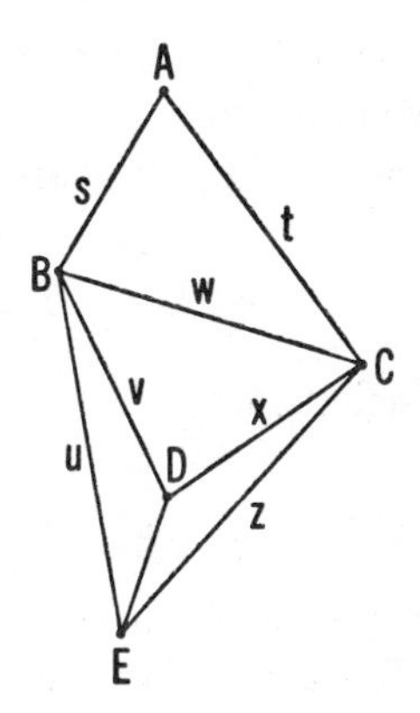

Figure 7.7

The law of conservation of current states that, at each vertex of the graph, the total amount of current flowing in equals the total amount of current flowing out. For example, the vertex C (corresponding to the horizontal segment at level c) receives current $tI + wI$ through wires AC and BC (corresponding to squares t and w) and C loses current $xI + zI$ through wires CD and CE (corresponding to squares x and z). The law cited above yields the relation

$$t + w = x + z$$

which is just one of our horizontal compatibility relations.

Since current flows in at the upper edge of the plate, moves down vertically and flows out at the lower edge, we see that every point at the same horizontal level is at the same potential, and points at higher horizontal levels have higher potentials than points at lower horizontal levels. According to Ohm's law, *the difference in potential between two points of the network joined by a wire is equal to the product of the current through that wire and the resistance offered by it.* For example, the potential difference between B and C (representing levels b and c of the plate) is $wI \cdot R$ (wI being the current entering at the upper edge of the square w, and R the resistance of any square; see p. 52). Moreover, *the potential difference is additive*; that is, if A and D are two vertices in the network connected by a path AB, BD through B, then the difference in potentials between A and D is the sum of the differences in potentials between A, B and B, D: $sI \cdot R + vI \cdot R$ in our illustration. As a consequence, for any closed path in the network, say $ABDCA$, the corresponding sum is zero†, and for any two paths from the same beginning point to the same end point, the corresponding sums are the same. Thus for ABD and ACD, we have

$$sI \cdot R + vI \cdot R = tI \cdot R + xI \cdot R$$

which reduces to

$$s + v = t + x.$$

This is one of our vertical compatibility relations.

The law of conservation of current and Ohm's law yield a system of linear equations equivalent to the system S of compatibility relations. Now it seems plausible that if a difference in potential is prescribed for a network such as ours (say the potentials at vertices A and E are given, or equivalently, at the top and bottom edges of the plate) then the current flowing in each wire is determined. This, in fact, is the content of the celebrated theorem by Kirchhoff:

If a potential difference between any two points of a network is prescribed, then the laws of conservation and Ohm determine uniquely the flow in each wire.

Since these physical laws are equivalent with the geometrical compatibility relations (the prescribed potential difference corresponding to the vertical dimension of the rectangle), the assertion that the system S has a unique solution (see p. 52) is seen to be just a corollary of Kirchhoff's theorem.

A rigorous proof of Kirchhoff's theorem is beyond the scope of this essay.‡

Let us now re-examine the problem of partitioning a rectangle into squares.

First we note that a solution of the linear system S may include non-positive quantities. These cannot be made to correspond to

† A path in our network leads from a vertex along an edge *in the direction* of a next vertex, etc. We take these directions into account by using signed quantities in this sum. For example, AC corresponds to tI, and CA corresponds to $-tI$.

‡ The interested reader may look it up, for example, in R. E. Scott, *Elements of Linear Circuits*, Addison-Wesley, 1965, Reading, Mass.

lengths of sides of subdivision squares, so such solutions of the system S are not solutions of the tiling problem.

Secondly, we recall that, in solving a system of linear equations (say by successively eliminating unknowns, or by using determinants, or in any other way), we use only rational operations, that is, addition, subtraction, multiplication, division. Thus, if all the coefficients in S are rational (and this is the case if the prescribed dimension of the rectangle—or equivalently the prescribed potential difference of the network—is rational), then all quantities in the solution of S are also rational. This implies that *a rectangle whose dimensions L and W are incommensurable* (*L/W irrational*) *cannot be tiled with squares.*

We might try to attack the rectangle-tiling problem by first devising a systematic method of enumerating all possible partitions into n squares and letting the integer n increase. We might then solve the system S for each partition and discard all partitions with geometrically non-feasible solutions. Even among the feasible solutions, many would be uninteresting (i.e., partitions into equal squares, etc.). Requirements such as (a) no two squares in the partition may be congruent, or (b) the partitioned rectangle may not contain a smaller rectangle, impose further conditions on the solutions of S. Perhaps we can approach the problem simultaneously from the other direction; that is, weed out all networks about which we know a priori that the solutions of the associated system S are either geometrically not feasible or produce unwanted partitions of types (a) or (b) above.†

A squaring or tiling in which no two squares are the same size is said to be "perfect". Tutte and his friends were chiefly after perfect tilings. They wanted to find a "perfect square", that is, a square with a perfect tiling. All their results involved non-square rectangles. And so they began to think that a perfect square did not exist. However, in 1939 Roland Sprague of Berlin found one, and, since then, many others have been discovered.

Attention was then turned to finding the perfect square with the smallest number of squares in it (i.e., of lowest order). The record to date is one of order 24, found by an amateur mathematician, T. H. Willcocks of Bristol, England. (See Figure 7.8.)

† Some headway in this direction has been made, for example, by applying Euler's formula to the graphs of our networks.

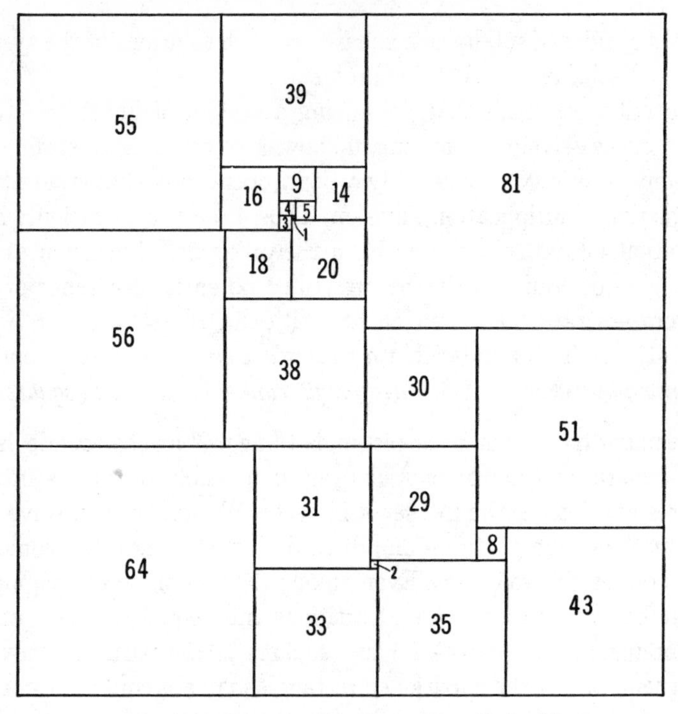

Figure 7.8. (Compound) perfect square of order 24.

A tiling is said to be "simple" if the arrangement of the squares does not form any smaller squared rectangle inside the original. Willcocks' perfect square is compound, not simple. Accordingly, attention was focussed on the *simple perfect square of lowest order*. Until recently, Willcocks also held this record with a square of order 37. However, in 1964 Dr. John Wilson of the University of Waterloo (a student of Tutte), using an electronic computer, found one of order 25 (see Figure 7.9), and he is the record-holder at present.

It has been shown, using computers, that there is no simple perfect square of order less than 20. Consequently, there is not much room for improvement in the above records.

So many perfect squares have been found that experts in the field suspect that tiled rectangles of unequal sides are the harder to come by.

We close this essay with the following easy-to-prove theorem: *It is impossible to fill a rectangular box with a finite number of unequal cubes.*

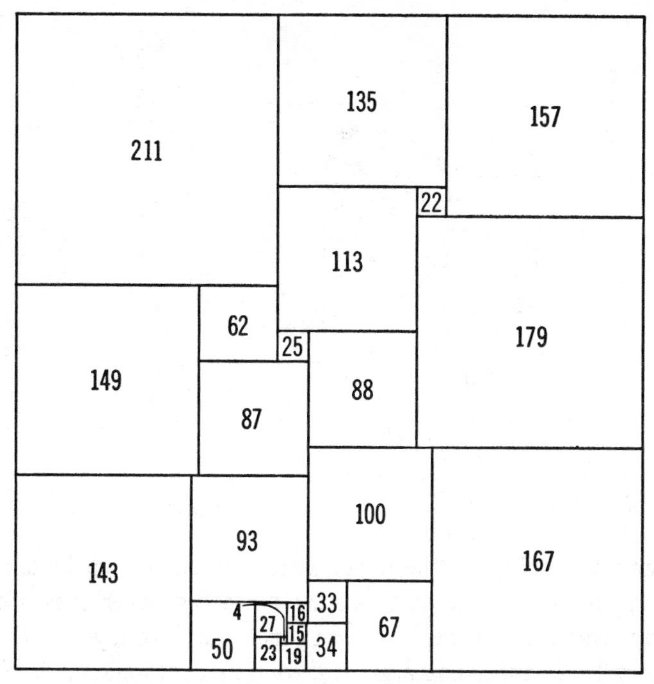

Figure 7.9. Simple perfect square of order 25.

Proof: Any successful packing of the box provides a perfect tiling of the bottom of the box by those cubes which rest on the bottom. Now the smallest cube S among those lining the bottom certainly could not touch an upright side of the box; for then there would have to be an even smaller one touching the bottom. (See Figure 7.10.)

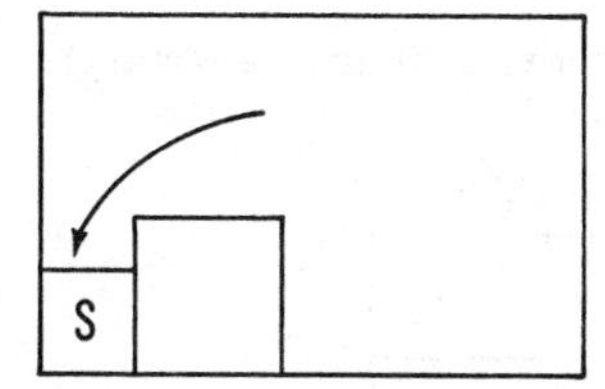

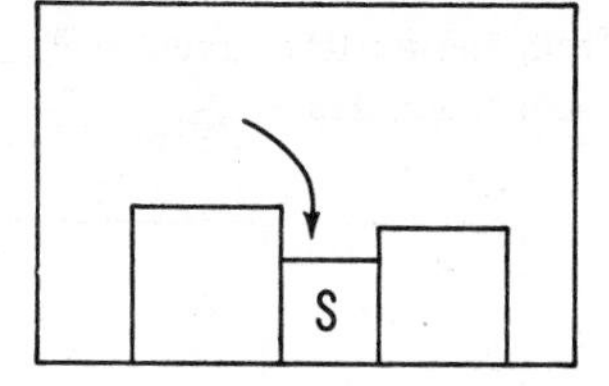

Figure 7.10. View of bottom rectangle of the box.

This smallest cube on the bottom, being out in the middle part of the bottom, must be bordered on every side by a bigger cube (it is the smallest). Its upper surface, then, is completely walled in; see Figure 7.11. In order to cover this upper surface, even smaller cubes must be used.

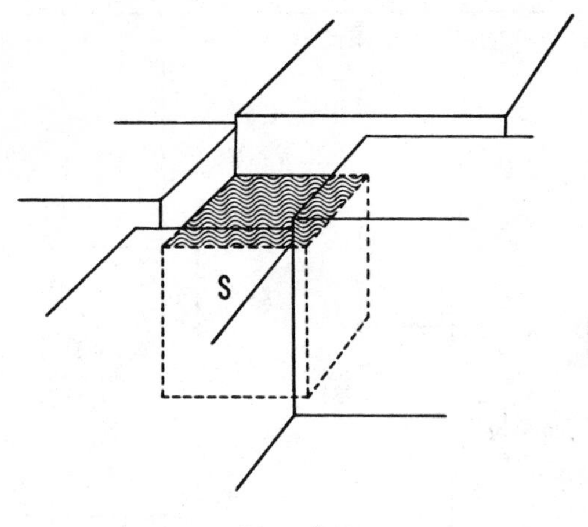

Figure 7.11

Among the cubes on the upper surface of S, the smallest, again occurring in the middle part, is surrounded by bigger ones. Consequently, even smaller ones still must occur in a third layer on the top of this walled-in cube. This argument continues without end, implying that there is no end to the number of cubes that must be employed.

EXERCISES

1. From the pattern given in Figure 7.12, derive a tiling of a rectangle into unequal squares.

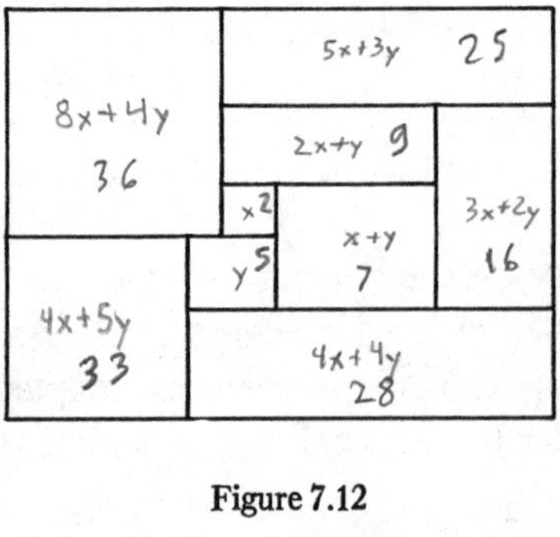

Figure 7.12

2. Prove that there exists some rectangle which can be tiled into N unequal squares for each value of N greater than 8, i.e., for $N = 9, 10, 11, \cdots$.

3. In tiling an equilateral triangle with unequal equilateral triangles, prove:

   (a) the smallest tile, S, which touches the base of the original triangle meets the base in just one point (see Figure 7.13).

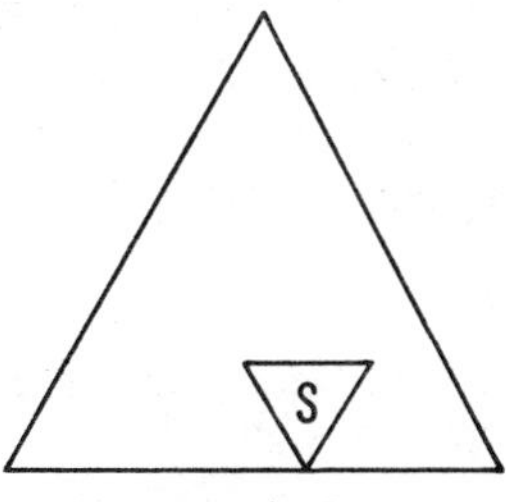

Figure 7.13

   (b) the smallest tile, T, touching the top of S meets the top of S in just one point (see Figure 7.14).

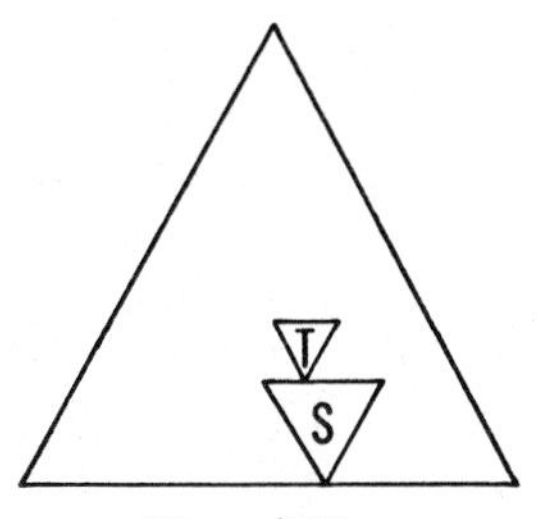

Figure 7.14

   (c) Does this lead to an endless tower of little triangles, providing a proof of the theorem "it is impossible to tile an equilateral triangle with unequal equilateral triangles (no two alike)", which was first proved by Tutte around 1948?

REFERENCES

C. J. Bouwkamp, *On the Dissection of Rectangles into Squares*, Kon. Ned. Ak. van Wet., 1946, Vol. 49, pp. 1176–1188; 1947, Vol. 50, pp. 1296–1299.

R. Brooks, C. Smith, A. Stone, W. Tutte, *The Dissection of Rectangles into Squares*, Duke Mathematics Journal, 1940, Vol. 7, pp. 312–340.

M. Gardner, Scientific American Magazine, November, 1958, pp. 136–142.

H. Meschkowski, *Unsolved and Unsolvable Problems in Geometry*, Oliver and Boyd, 1966, Edinburgh, pp. 91–102.

S. Stein, *Mathematics: The Man-Made Universe*, (2nd ed.) Freeman and Co., 1969, San Francisco, pp. 92–124.

W. Tutte, *Squaring the Square*, Canadian Journal of Mathematics, 1950, pp. 197–209.

W. Tutte, *The Quest of the Perfect Square*, The American Mathematical Monthly, 1965, Vol. 72, No. 2, pp. 29–35.

ESSAY EIGHT

Writing a Number as a Sum of Two Squares

In the theory of numbers there is an arithmetic function, $r(n)$, defined as the number of ways in which the non-negative integer n can be expressed as the sum of two squares of integers. More precisely, $r(n)$ is the number of ordered pairs (x, y) such that $n = x^2 + y^2$. For example, $r(5) = 8$ because

$$\begin{aligned} 5 &= (+1)^2 + (+2)^2 = (+2)^2 + (+1)^2 \\ &= (-1)^2 + (+2)^2 = (+2)^2 + (-1)^2 \\ &= (+1)^2 + (-2)^2 = (-2)^2 + (+1)^2 \\ &= (-1)^2 + (-2)^2 = (-2)^2 + (-1)^2; \end{aligned}$$

these expressions correspond, respectively, to the ordered pairs

$$(1,2), (2,1), (-1,2), (2,-1), (1,-2), (-2,1), (-1,-2), (-2,-1).$$

A few values of this function are

$$\begin{aligned} &r(0) = 1, \quad r(1) = 4, \quad r(2) = 4, \quad r(3) = 0, \\ &r(4) = 4, \quad r(5) = 8, \quad r(7) = 0, \quad r(12) = 0. \end{aligned}$$

There are various facts that can be proved about $r(n)$. For example, $r(n) = 0$ if n is of the form $4k + 3$ (i.e., if n is any number in the arithmetic progression $3, 7, 11, 15, 19, 23, \cdots$). Consequently the value of the function drops to 0 infinitely often as n runs through the non-negative integers $0, 1, 2, 3, \cdots$. But the function $r(n)$ takes on large values too. In fact, it is not hard to find an n which will make $r(n)$ as big as you want. It is an extremely irregular function. In such a situation, it is a good idea to look at the average value of $r(n)$ over a set of consecutive values of n. This is where the function is surprisingly well-behaved.

The average value of $r(n)$ over the integers $0, 1, 2, \cdots, z-1$ is

$$\frac{r(0) + r(1) + r(2) + \cdots + r(z-1)}{z}$$

If we call the numerator of this fraction $R(z)$, we may denote this average value more briefly by $R(z)/z$. The average value of $r(n)$ over the entire range of non-negative integers is defined to be the limit of this expression as z increases indefinitely,

$$\lim_{z\to\infty} [R(z)/z],$$

provided the limit exists. It turns out that *this limit exists, and is equal to* π! On the average, then, a non-negative integer has π representations as the sum of the squares of two integers. One might well wonder how the number π could have any relation to $r(n)$. And it is surprising that the proof of the above result is neither sophisticated nor subtle. It is highly instructive and very interesting. The German mathematician Gauss (1777–1855) found it around 1800, when he was 23 years old.

Proof: In the Cartesian coordinate plane, consider the circle

$$C(\sqrt{z}): \quad x^2 + y^2 = z$$

with centre at the origin and radius $\sqrt{z}$. By a *lattice point* $A = (a, b)$ we mean a point whose coordinates are integers. Every lattice point A inside $C(\sqrt{z})$ has coordinates satisfying the inequality $a^2 + b^2 < z$, since the distance from A to the origin is less than the radius $\sqrt{z}$

of $C(\sqrt{z})$, that is,

$$\sqrt{a^2 + b^2} < \sqrt{z},$$

(see Figure 8.1). Moreover, since a and b are integers, $a^2 + b^2 = n$ is an integer. Thus, the ordered pair (a, b) leads to an expression for n as the sum of the squares of two integers:

$$a^2 + b^2 = n < z.$$

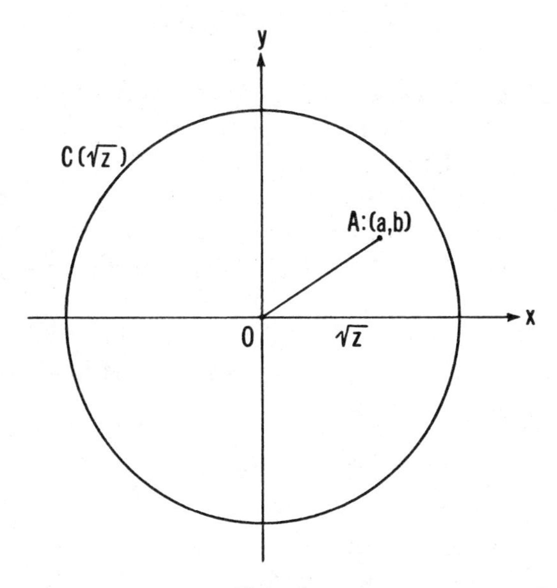

Figure 8.1

Every lattice point inside $C(\sqrt{z})$, then, contributes 1 to the sum $R(z) = r(0) + r(1) + \cdots + r(z - 1)$, since it furnishes an ordered pair counted by some $r(n)$ in $R(z)$. Conversely, any ordered pair (p, q), such that

$$p^2 + q^2 = n < z,$$

i.e., any ordered pair counted by $R(z)$, must be the coordinates of a lattice point inside $C(\sqrt{z})$. Consequently, $R(z)$ *is equal to the number of lattice points in* $C(\sqrt{z})$. Now we investigate how many lattice points there are inside $C(\sqrt{z})$.

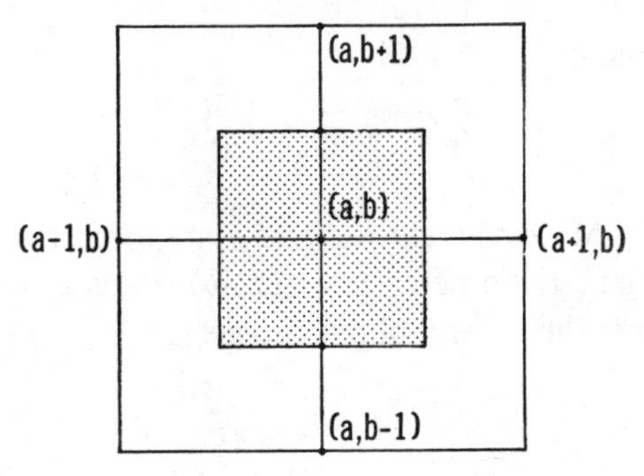

Figure 8.2

Around each lattice point $P = (a, b)$ in the plane let us place a square of side 1 centred at P (see Figure 8.2). We colour blue the squares around the lattice points inside $C(\sqrt{z})$ and red all other squares. As a result, most of the interior of $C(\sqrt{z})$ is blue and most of its exterior is red. However, some blue squares project beyond $C(\sqrt{z})$ and some red ones cut into $C(\sqrt{z})$ (see Figure 8.3). Now, the number of blue squares is the number of lattice points in $C(\sqrt{z})$, and this is equal to $R(z)$. Also, since each square has area 1, the number of blue squares is the total area coloured blue, which we denote by A_b. Thus

$$R(z) = A_b.$$

Now we estimate A_b.

If Q is a lattice point not in the interior of $C(\sqrt{z})$, and if R is any point belonging to the unit square with centre Q (Figure 8.4), then

$$OQ \geq \sqrt{z}, \qquad RQ \leq 1/\sqrt{2},$$

and by the triangle inequality, $OR + RQ \geq OQ$, we have

$$OR \geq OQ - RQ \geq \sqrt{z} - 1/\sqrt{2}.$$

It follows that no red-coloured point lies inside the circle $C(\sqrt{z} - 1/\sqrt{2})$ with centre at the origin and radius $\sqrt{z} - 1/\sqrt{2}$. Similarly, for a blue-coloured point B belonging to a square with interior lattice point P as centre, we have $OP < \sqrt{z}$, $PB \leq 1/\sqrt{2}$, and

$$OB \leq OP + PB < \sqrt{z} + 1/\sqrt{2},$$

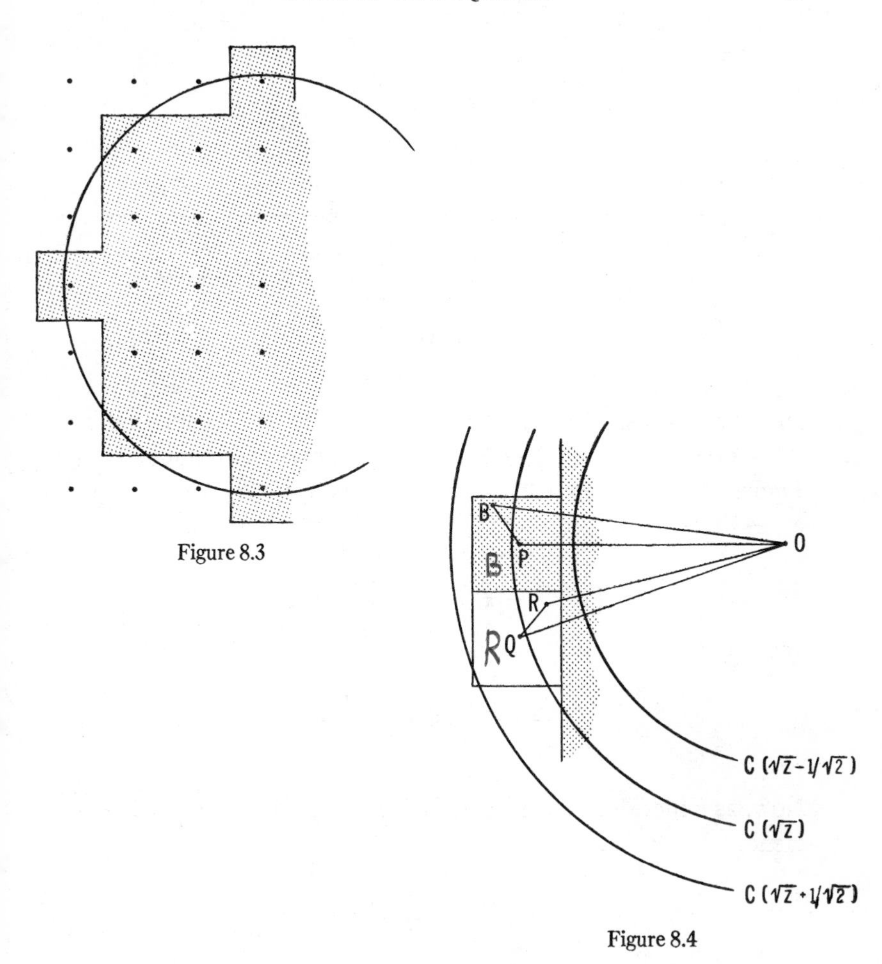

Figure 8.3

Figure 8.4

so no blue-coloured point lies outside the circle $C(\sqrt{z} + 1/\sqrt{2})$. It follows that area A_b lies between the areas of the circles $C(\sqrt{z} - 1/\sqrt{2})$ and $C(\sqrt{z} + 1/\sqrt{2})$:

$$\text{Area } C(\sqrt{z} - 1\sqrt{2}) \leq A_b = R(z) \leq \text{Area } C(\sqrt{z} + 1/\sqrt{2}),$$

that is,

$$\pi(\sqrt{z} - 1/\sqrt{2})^2 \leq R(z) \leq \pi(\sqrt{z} + 1/\sqrt{2})^2,$$

$$\pi(z - \sqrt{2z} + 1/2) \leq R(z) \leq \pi(z + \sqrt{2z} + 1/2),$$

$$\pi z - \pi\sqrt{2z} + \pi/2 \leq R(z) \leq \pi z + \pi\sqrt{2z} + \pi/2,$$

whence

$$\pi/2 - \pi\sqrt{2z} \leq R(z) - \pi z \leq \pi/2 + \pi\sqrt{2z}.$$

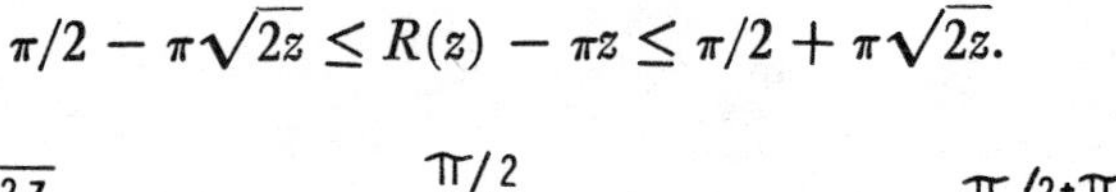
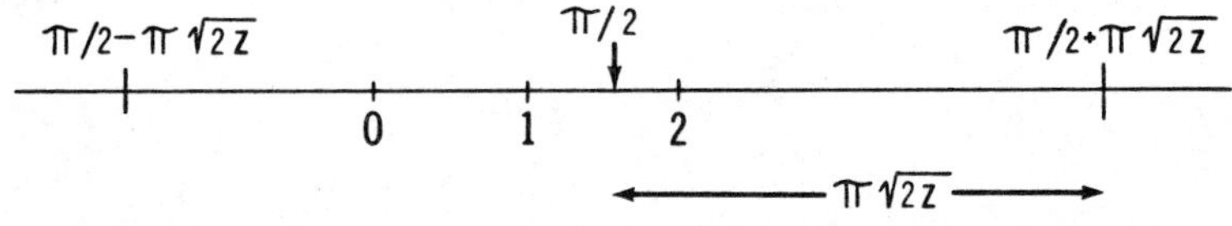

Figure 8.5

In other words, the number $R(z) - \pi z$ is in the interval from $\pi/2 - \pi\sqrt{2z}$ to $\pi/2 + \pi\sqrt{2z}$, including the endpoints (see Figure 8.5). Consequently, the magnitude of $R(z) - \pi z$ cannot exceed $\pi/2 + \pi\sqrt{2z}$; that is,

$$|R(z) - \pi z| \leq \pi(\sqrt{2z} + 1/2),$$

and

$$\left|\frac{R(z)}{z} - \pi\right| \leq \pi\left(\sqrt{\frac{2}{z}} + \frac{1}{2z}\right).$$

Now, as z increases, the second factor on the right approaches zero. Hence

$$\lim_{z\to\infty}\left|\frac{R(z)}{z} - \pi\right| = 0,$$

which is equivalent to the statement

$$\lim_{z\to\infty}\frac{R(z)}{z} = \pi.$$

REFERENCE

D. Hilbert and S. Cohn–Vossen, *Geometry and the Imagination*, Chelsea, 1952, New York.

ESSAY NINE

The Isoperimetric Problem

The isoperimetric problem is to find, among all closed curves in the plane of given perimeter p, the one that bounds a region with the greatest possible area. Speaking loosely, how should a loop of string, p units around, be arranged on a flat surface so as to enclose the largest possible area?

The reader might guess that a circle is the right answer. However, this requires a proof.

In the first place, it is conceivable that there might be no shape which gives a maximum area; that is, that no matter what shape the loop is given, there exists a loop of another shape and the same length p enclosing a bigger area. This is a sticky problem which we shall not consider here.† We proceed to determine the shape of the maximum area, on the assumption that it does exist. The work presented below is that of the Swiss-German mathematician Jacob Steiner (1796–1863). Steiner advances in three steps.

(a) Firstly, he shows that the required curve (which, hereafter, we shall call $\mathcal{C}$) must enclose a convex region; i.e., a region with no humps or dips. (More precisely: denote by G the closed region bounded by the required curve $\mathcal{C}$; G is called *convex* if, for every pair of points P_1, P_2 of G, the whole line segment connecting P_1 and P_2 lies in G.)

† For a convincing argument that this isoperimetric problem has a solution, see, e.g., N. Kazarinoff, *Geometric Inequalities*, (NML 4), Random House, 1961, New York, pp. 58–63.

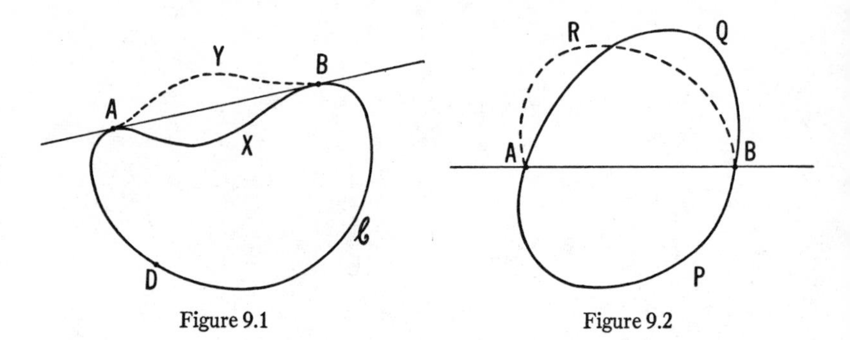

Figure 9.1

Figure 9.2

We see this by noting that a curve bounding a non-convex region cannot solve our problem. For, any such curve has at least one pair of points A, B (Figure 9.1) such that the entire curve lies on one side of the line through A and B. If the dip AXB is reflected in this line we obtain a curve $DAYB$ which, clearly, encloses a greater area than our original curve, but which has the same perimeter (arc AXB has the same length as its mirror image AYB). Hence G must be a convex region.

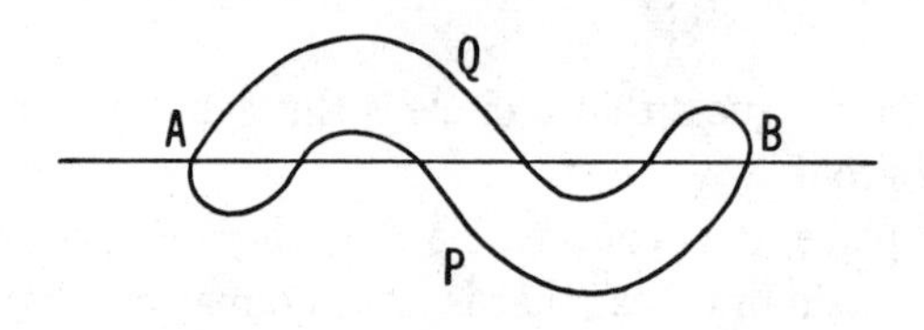

Figure 9.3

(b) Secondly, Steiner shows that any straight line which divides the perimeter of G in half also divides its area in half. Let $APBQ$ represent the boundary $\mathfrak{C}$ of G and let AB divide the perimeter in half. Since G is convex, each half of the bounding curve lies entirely on one side of the line through A and B (see Figure 9.2).† Suppose that the parts of G on opposite sides of AB are unequal in area, APB (say) being the larger. Then, reflecting APB in AB as a mirror, we obtain a curve $APBR$ whose perimeter is the same as the perimeter of G, but which bounds a bigger area (twice area APB is greater than area APB + area AQB.). This contradiction establishes that AB divides the area of G in half.

† Note that this argument is valid only if G is convex (see Figure 9.3).

(c) We learn from (b) that, running from *any* point A on $\mathcal{C}$ halfway around to a point B, we obtain a region AQB, enclosed by the arc AQB and the straight line AB, whose area is one-half the area of G. Because G is convex, this region AQB lies entirely on one side of AB. Now we ask the question: Among all curves of perimeter one-half p (i.e., one-half the given perimeter) which enclose convex regions on one side of a straight line and have their endpoints on that straight line, is there one containing an area greater than one-half G? The answer is no. For, if there were such a curve, this curve together with its mirror image in "its line" would give a (full) curve of perimeter p enclosing an area greater than that of G; and this is impossible. Hence we may conclude: Any arc AB of length one-half p provides us with a figure AQB which contains the largest area that can be enclosed by a curve of length one-half p and a straight line. In part (c), we direct our attention to any such configuration AQB.

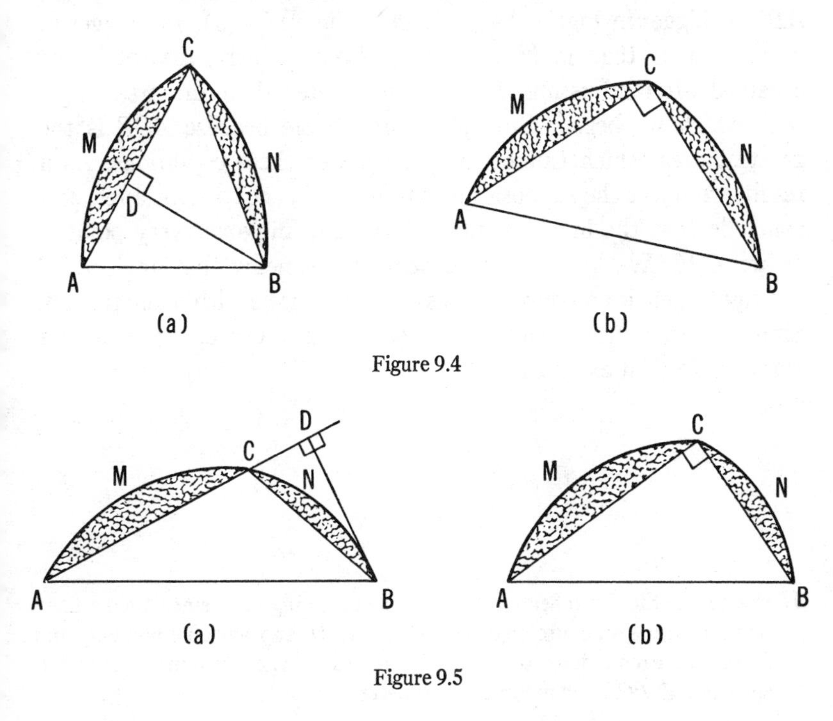

Figure 9.4

Figure 9.5

Select any point C on this "half curve" AQB, and join it to A and to B, thus obtaining two lunes, AMC and BNC, and a triangle ABC (Figure 9.4). The angle C of the triangle is either acute, obtuse or a right angle. We shall show that it is a right angle by eliminating the other possibilities. Let us imagine that the lunes are made of non-deformable material and that they are hinged at C like a pair of crab claws. Now suppose the angle at C were acute [as in Figure 9.4(a)] or obtuse [as in Figure 9.5(a)]. Then by opening or closing the lunes, respectively [see Figures 9.4(b) and 9.5(b)], we can obtain a right angle at the hinge C. This operation does not alter the perimeters or areas of the non-deformable lunes; the only change occurs in the triangle ABC. Its area may be determined, in all cases, by using AC as base and BD as height. [D coincides with C in Figures 9.4(b) and 9.5(b)]. Since the length of AC is constant, a bigger area is obtained from a bigger altitude. In Figures 9.4(a) and 9.5(a) the altitude BD, a leg of the right-angled triangle BCD, is clearly less than the hypotenuse BC. But BC, which does not change in moving the lunes, is the altitude in Figures 9.4(b) and 9.5(b); that is, triangle ABC is bigger in Figure 9.4(b) than in Figure 9.4(a), and bigger in Figure 9.5(b) than in Figure 9.5(a). Consequently, the total area obtained after adjusting the lunes is greater than the area AQB with which we began. But this is impossible because AQB is the greatest area which can be enclosed under these conditions. As a result, we reject the supposition that angle C is not a right angle and conclude that the line AB subtends a right angle at every point C of arc AQB. We know from elementary geometry that angle ACB is a right angle if and only if C lies on a semi-circle with diameter AB. Similarly, the "other" half of $\mathcal{C}$ must be a semi-circle, implying that $\mathcal{C}$ is a circle (if it exists at all).

EXERCISES

1. Use the method of reflection to solve the following problem: Given a fixed point P inside an acute angle α with vertex O, and variable points Q and R, one on each side of the angle α. Determine the positions of Q and R such that $\triangle PQR$ has minimum perimeter.

2. Find the other two sides of the triangle with base 5 units and area 15 units which has the smallest possible perimeter.

REFERENCE

R. Courant and H. Robbins, *What is Mathematics?*, (4th ed.), Oxford University Press, 1947, New York.

ESSAY TEN

Five Curiosities from Arithmetic

In this section we shall take up five little gems of an arithmetic nature. We describe them all before delving into their proofs.

I. Liouville's Generalization

It is well known that

$$(1) \qquad 1^3 + 2^3 + 3^3 + \cdots + n^3 = (1 + 2 + 3 + \cdots + n)^2$$

for all positive integers n (see Exercise 1). The French mathematician Joseph Liouville (1809–1882) discovered a surprising procedure for producing other sets of positive integers with the same property: *the sum of their cubes equals the square of their sum.*

First choose a positive integer N, say 6. Next determine the divisors of N; for $N = 6$, they are $(1, 2, 3, 6)$. Finally, determine the *number* of divisors of these divisors—in the case at hand $(1, 2, 2, 4)$—and you will have a set of numbers with the desired property:

$$1^3 + 2^3 + 2^3 + 4^3 = 1 + 8 + 8 + 64 = 81 = 9^2 = (1 + 2 + 2 + 4)^2.$$

II. The Number 6174

Given the number 6174, arrange its digits so as to form the largest number possible from them, i.e., put them in decreasing order. Also arrange them to form the smallest number possible, and subtract. We get

$$7641 - 1467 = 6174,$$

the number we began with.

Let us apply this same procedure to the number 4959. We get

$$9954 - 4599 = 5355.$$

There doesn't seem to be anything specially noteworthy here. Let's carry on by applying the procedure to the difference 5355. We obtain

$$5553 - 3555 = 1998;$$

again nothing special. Nevertheless, let us continue to apply our procedure to succeeding results. We get, in turn,

$$\begin{aligned} 9981 - 1899 &= 8082, \\ 8820 - 0288 &= 8532, \\ 8532 - 2358 &= 6174. \quad \text{Aha!} \end{aligned}$$

The fact is that *no matter what four-digit number one starts out with, provided its digits are not all the same, this procedure turns up the number* 6174 *in at most* 7 *steps*.

III. Professor Ducci's Observation

In the 1930's the following observation was attributed to Professor E. Ducci of Italy.

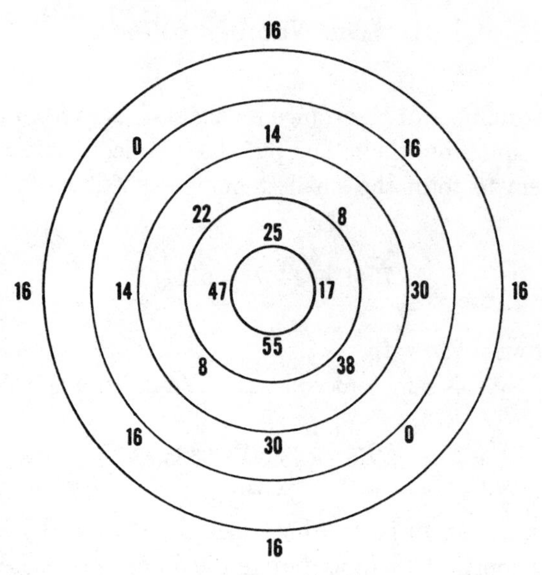

Figure 10.1

Put any four non-negative integers around a circle, say 25, 17, 55, and 47. (See Figure 10.1.) Now construct further "cyclic quadruples" of integers by subtracting consecutive pairs, always the smaller from the bigger. Eventually, *four equal numbers will occur*.

IV. The Sum of the Squares of the Digits of a Number

Begin with any positive integer, say 9246, and determine the sum of the squares of its digits $(81 + 4 + 16 + 36 = 137)$. Do the same for this number (137 gives $1 + 9 + 49 = 59$), and with each succeeding result, to obtain a sequence of integers. For our example the sequence is

$$9246, 137, 59, 106, 37, 58, 89, 145, 42, 20, \cdots.$$

Then, no matter what integer one chooses initially, *the resulting sequence either heads for the number* 1 (*after which it clearly repeats indefinitely*), *or it heads for the number* 4 (*at which point the cycle* 4, 16, 37, 58, 89, 145, 42, 20 *repeats indefinitely*).

V. Sundaram's Sieve

Many readers are probably acquainted with the "sieve of Eratosthenes" for filtering prime numbers. In 1934, a young east Indian student named Sundaram proposed the following alternative.

In Table 10.1, the first row is composed of the terms of the arithmetic progression 4, 7, 10, $\cdots$, with first term 4 and common difference 3. This progression is also used to provide the first column. Succeeding rows are then completed so that each consists of an arithmetic progression, the common differences in successive rows being the odd integers 3, 5, 7, 9, 11, $\cdots$.

4	7	10	13	16	19	22	25	$\cdots$
7	12	17	22	27	32	37	42	$\cdots$
10	17	24	31	38	45	52	59	$\cdots$
13	22	31	40	49	58	67	76	$\cdots$
16	27	38	49	60	71	82	93	$\cdots$
$\cdot$	$\cdot$	$\cdot$	$\cdot$	$\cdot$	$\cdot$	$\cdot$	$\cdot$	$\cdots$

Table 10.1

The remarkable property of Table 10.1 is: *if N occurs in the table, then $2N + 1$ is not a prime number; if N does not occur in the table, then $2N + 1$ is a prime number.*

We now give the proofs. In each case we first assert the principle we shall prove.

I. Let N be any integer and $(d_1, d_2, \cdots, d_k)$ the set of all its divisors (including 1 and N itself). Let c_j be the number of divisors of d_j. Then the set $(c_1, c_2, \cdots, c_k)$ has the property

$$c_1^3 + c_2^3 + \cdots + c_k^3 = (c_1 + c_2 + \cdots + c_k)^2.$$

Proof: (a) It is easy to see that Liouville's theorem holds for a power of a prime, $N = p^n$. In this case the divisors are

$$p^0, \; p^1, \; p^2, \; \cdots, \; p^n.$$

By a direct count we see that the number of divisors of p^n is $n + 1$.

This holds for any n, hence the numbers of divisors of the above divisors are

$$1, 2, 3, \cdots, n+1,$$

and the property holds for this set of numbers by relation (1) on page 72 applied to the first $n+1$ integers.

(b) Next we proceed inductively by showing that if the theorem is valid for the positive integer K, then it is valid also for $N = Kp^n$, where p is a prime not dividing K. When this is established, the conclusion easily follows. For, if

$$N = p_1^{a_1} \cdot p_2^{a_2} \cdot p_3^{a_3} \cdots p_k^{a_k}$$

is the prime factorization of N, we know by (a) that the theorem holds for

$$N_1 = p_1^{a_1},$$

and by (b) that it holds also for

$$N_2 = N_1 \cdot p_2^{a_2} = p_1^{a_1} \cdot p_2^{a_2},$$

and then for

$$N_3 = N_2 \cdot p_3^{a_3} = p_1^{a_1} \cdot p_2^{a_2} \cdot p_3^{a_3},$$

etc., to

$$N = p_1^{a_1} \cdot p_2^{a_2} \cdots p_k^{a_k},$$

itself. It only remains, then, to prove (b).

Let us denote the divisors of K by $A, B, \cdots, K$,† and the number of divisors of these divisors, respectively, by $a, b, \cdots, k$. Because we are supposing that the procedure holds for K, we have

$$a^3 + b^3 + \cdots + k^3 = (a + b + \cdots + k)^2.$$

Clearly, any divisor of K multiplied by any divisor of p^n yields a divisor of Kp^n. Here is a complete list of such products:

† Every number is a divisor of itself; in our notation this is the last divisor listed.

$$(2)\quad \begin{gathered} A,\ B,\ \cdots,\ K;\ \ pA,\ pB,\ \cdots,\ pK;\ \ p^2A,\ p^2B,\ \cdots,\ p^2K; \\ \cdots;\ \ p^nA,\ p^nB,\ \cdots,\ p^nK. \end{gathered}$$

Conversely, all divisors of Kp^n are of this form. For, let T be a divisor of Kp^n. T can be factored uniquely as

$$T = tp^m$$

where t is relatively prime to p;† and, since p does not divide K, $m \leq n$. So p^m divides p^n, and t divides K.

Moreover, all the divisors in the list (2) are distinct. Any two, say p^iG and p^jH either have different first factors (i.e., $i \neq j$) or they have different second factors (i.e., $G \neq H$), or both. Since p does not divide K, it follows that neither G nor H contains p as factor, so $p^iG \neq p^jH$.

The next thing we need to know is how many divisors a number like p^iA has. As in the case of Kp^n, the divisors of p^iA are the $i + 1$ divisors of p^i taken in combination with the a divisors of A. Clearly, then, there are $(i + 1)a$ of them. Accordingly, the final set of number of divisors of the divisors of $N = Kp^n$ is

$$\begin{gathered} a,\ b,\ \cdots,\ k;\ \ 2a,\ 2b,\ \cdots,\ 2k;\ \ 3a,\ 3b,\ \cdots,\ 3k; \\ \cdots;\ \ (n+1)a,\ (n+1)b,\ \cdots,\ (n+1)k. \end{gathered}$$

The sum of these numbers easily simplifies to

$$S = (a + b + \cdots + k)[1 + 2 + \cdots + (n + 1)].$$

Hence the square of their sum is

$$\begin{aligned} S^2 &= (a + b + \cdots + k)^2[1 + 2 + \cdots + (n + 1)]^2 \\ &= (a^3 + b^3 + \cdots + k^3)[1^3 + 2^3 + \cdots + (n + 1)^3], \end{aligned}$$

since the property in question holds for the numbers in each bracket. Thus

† This is a consequence of the unique factorization theorem.

$$\begin{aligned} S^2 &= (a^3 + b^3 + \cdots + k^3) + (2^3a^3 + 2^3b^3 + \cdots + 2^3k^3) \\ &\quad + \cdots + [(n+1)^3a^3 + (n+1)^3b^3 + \cdots + (n+1)^3k^3] \\ &= a^3 + b^3 + \cdots + k^3 + (2a)^3 + (2b)^3 + \cdots + (2k)^3 \\ &\quad + \cdots + [(n+1)a]^3 + [(n+1)b]^3 + \cdots + [(n+1)k]^3, \end{aligned}$$

as required.

II. Let M be a four digit number with not all four digits the same. Form numbers M_l and M_s by arranging the digits of M in decreasing order, then in increasing order, respectively, and compute the difference $D_1 = M_l - M_s$.

By this method we assign to every permissible four digit number M a four-digit number D_1.† This assignment, or *transformation*, maps every number M into a number D_1 and may be denoted by T. We say

$$T\colon\ M \to D_1 \qquad \text{or} \qquad T(M) = D_1.$$

We must show that at most seven repeated applications of the transformation T produces the number 6174; i.e.,

$$T(M) = D_1,\ T^2(M) = T(D_1) = D_2,\ \cdots,\ T^k(M) = D_k = 6174$$

for $k \leq 7$.

Proof: There are $10^4 = 10000$ four digit numbers;† but those with four equal digits are clearly mapped into 0000 by our transformation T. There remain $10^4 - 10 = 9{,}990$ different four-digit numbers with not all digits the same. We shall show first that the transformation T maps these 9,990 numbers into only 54 four-digit numbers. For, let $a,\ b,\ c,\ d$ be the digits of M, named so that

$$a \geq b \geq c \geq d. \tag{3}$$

Since not all are equal, not all three equality signs in (3) can hold simultaneously. We calculate $T(M)$:

† Numbers whose first digit is 0, such as 0020 or 0999, are included among these four-digit numbers.

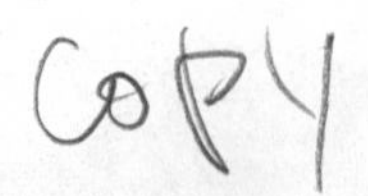

$$
\begin{aligned}
M_l &= 1000a + 100b + 10c + d, \\
M_s &= 1000d + 100c + 10b + a, \\
D_1 = M_l - M_s &= 1000(a-d) + 100(b-c) + 10(c-b) + (d-a) \\
= \quad T(M) &= 999(a-d) + 90(b-c).
\end{aligned}
$$

We note that $T(M)$ depends on $(a-d)$ and $(b-c)$. Since not all digits a, b, c, d are equal, the inequalities (3) yield

$$a - d > 0, \tag{4}$$

and $b - c \geq 0$. Moreover, b, c lie between a and d, so

$$a - d \geq b - c. \tag{5}$$

The inequalities (4) and (5) imply that $a - d$ may take the nine values $1, 2, \cdots, 9$, and if it takes some value n in this set, $b - c$ can take at most the values $0, 1, 2, \cdots, n$. For example, if $a - d = 1$, the only choices for $b - c$ are 0 and 1; consequently $T(M)$ may have only the values

$$
\begin{aligned}
999(1) + 90(0) &= 0999, \\
999(1) + 90(1) &= 1089
\end{aligned}
$$

in this case. Similarly, if $a - d = 2$, $T(M)$ can have only the three values corresponding to $b - c = 0, 1, 2$. Adding the number of possible values for $b - c$ in the cases $a - d = 1$, $a - d = 2$, $\cdots$, $a - d = 9$, we obtain

$$2 + 3 + 4 + \cdots + 10 = 54$$

possible values for $T(M)$.

Next we observe that two numbers M and N having the same digits, but not in the same order, have the same image under the transformation T: $T(M) = T(N)$. Let us say that two numbers are *equivalent* if they contain the same quadruple of digits. Among the 54 possible values for $T(M)$, only 30 are non-equivalent (see Exercise 3). These are listed in Table 10.2

9990	9981	9972	9963	9954	9810	9711	9621	9531	9441
8820	8730	8721	8640	8622	8550	8532	8442	7731	7641
7632	7551	7533	7443	6642	6552	6543	6444	5553	5544

Table 10.2

By pursuing each of these 30 cases individually (see Exercise 4) we see that 6174 is turned up in at most 6 more steps (QED).

Before leaving this section, we might note that for six-digit numbers there are 384 different results possible at the end of the first subtraction. Of these 30 lead to the "repeater" 631764, 353 lead to the cycle (840852, 860832, 862632, 642654, 420876, 851742, 750843), and in one case only does the repeater 549945 arise.

A quick look at the results for eight-digit numbers indicates that similar conclusions occur also in this case. Finally, the similarity of the six-digit 631764 to 6174 is so striking that one cannot resist trying 63317664, 6333176664, etc. (What happens?)

III. Four non-negative integers g, h, k, l are arranged in cyclic order; the absolute values of the four differences of adjacent integers,

$$|g-h|, \quad |h-k|, \quad |k-l|, \quad |l-g|,$$

are arranged in cyclic order to form the next "cyclic quadruple". Each cyclic quadruple consists of the absolute values of the differences of adjacent members of the previous cyclic quadruple. After a finite number of steps, the resulting quadruple consists of four equal integers.

Proof: If our initial quadruple has four equal integers, there is nothing to prove. If not all four are equal, we show that the biggest integer† must get smaller in four or fewer steps. For example, if the biggest number at some point is 30, then four steps later it cannot be more than 29, another four steps later not more than 28, etc.; in at most 120 steps altogether, then, the biggest number reaches zero.

† If two or three are equal and bigger than the remaining, each of these may be called "the biggest".

But if the biggest number is 0, all four numbers must be zeros, and before this can happen for the first time, the four numbers of the previous stage must all have been the same. All we have to show, then, is that the biggest number decreases in four or fewer stages.

Clearly the biggest number will get smaller right away if none of the four numbers is zero. If a zero is present, however, it could happen that the biggest number escapes to the next stage undiminished. We prove, then, that zeros are merely a temporary phenomenon; a quadruple containing either one, two or three zeros will become one with no zeros at all in three or fewer stages.

(i) *Three Zeros*. Figure 10.2 demonstrates our claim. All zeros have disappeared in three stages.

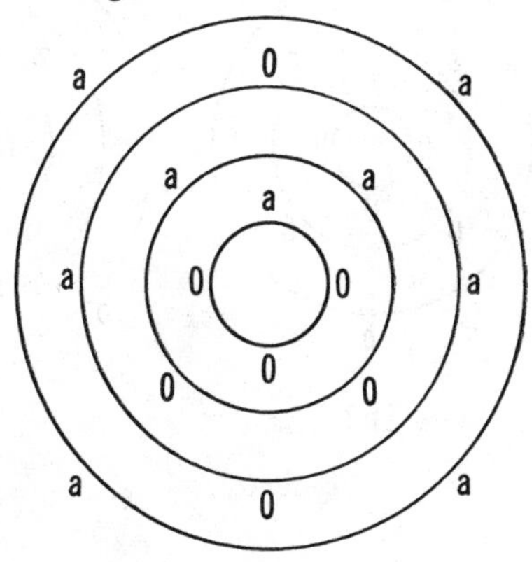

Figure 10.2

(ii) *Two Zeros*. Let a and b represent non-zero values.

(a) If the two zeros are opposite each other around the circle [see Figure 10.3(a)], both have disappeared in the next stage.

(b) If the two zeros are next to each other [see Figure 10.3(b)], consider the next two stages, whose quadruples we denote by $(0, a, c, b)$, (a, e, d, b). There are two cases: $c = 0$, $c \neq 0$.

(b_1) $c = 0$ occurs if the initial non-zero values a and b are equal, and this happened in the second stage of Figure 10.2. All zeros are gone after two steps.

(b_2) $c \neq 0$ occurs if $a \neq b$.

Case (b_2) breaks down into two sub-cases, $d = 0$ and $d \neq 0$, which occur when $b = c$ and when $b \neq c$, respectively.

In the first subcase, $c = |a - b| = b$, and the only solutions of this equation are $a = 0$ and $a = 2b$. Since a is not zero, we reject the first. In other words, we began with the quadruple $(2b, b, 0, 0)$. Figure 10.3(b′) shows that all zeros disappear in three stages.

In the subcase $d \neq 0$, we go on to consider e. But the role played by e is similar to that played by d; i.e., if $e = 0$, $a = c = |a - b|$, so that $b = 2a$, the original quadruple was $(a, 2a, 0, 0)$ and all zeros disappear in 3 stages.

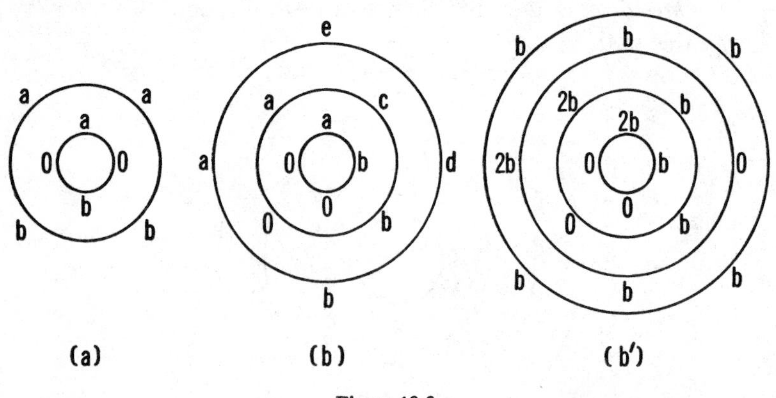

Figure 10.3

(iii) *One Zero.* If all three numbers are different, the zero disappears after one step. The same is true if diametrically opposite non-zero values are equal but different from the third.

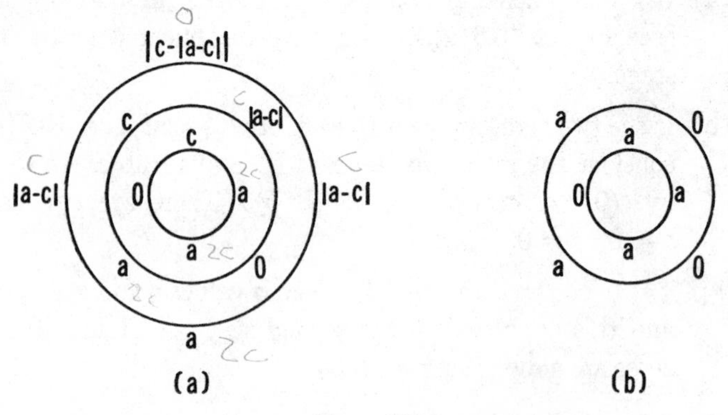

Figure 10.4

If two adjacent numbers are equal but different from the third, say $a = b$ and $a \neq c$, then no zeros are left after two steps provided that $|c - |a - c|| \neq 0$ [see Figure 10.4(a)], i.e. provided that $a \neq 2c$. If $a = 2c$, the outermost cycle in Figure 10.4(a) reduces to $0, c, 2c, c$ which yields the cycle of equal numbers c, c, c, c in the next step.

If $a = b = c$, no zeros appear after three steps [see Figure 10.4(b) and the second stage in Figure 10.2].

IV. Let N be a positive integer, let $N_1 = S(N)$ be the sum of the squares of its digits, $N_2 = S(N_1)$ the sum of the squares of the digits of $N_1, \cdots$, and $N_k = S(N_{k-1})$ the sum of the squares of the digits of N_{k-1}. Then the sequence

$$N,\ N_1,\ N_2,\ \cdots,\ N_k,\ \cdots$$

has the property that either all terms from some k on have the value 1, or some term, say N_l, has the value 4 so that the eight terms

$$4,\ 16,\ 37,\ 58,\ 89,\ 145,\ 42,\ 20$$

keep repeating from this point on.

Proof: We prove this assertion in two parts: (i) for numbers with three or more digits, (ii) for numbers with one or two digits.

(i) Let $N = a_n 10^n + a_{n-1}10^{n-1} + \cdots + a_1 10 + a_0$, $a_n \neq 0$, $n \geq 2$. We show that $N > S(N)$ by observing that the difference,

$$\begin{aligned} N - S(N) &= a_n 10^n + a_{n-1}10^{n-1} + \cdots + a_1 10 + a_0 \\ &\quad - a_n^2 - a^2_{n-1} - \cdots - a_1^2 - a_0^2 \\ &= a_n(10^n - a_n) + a_{n-1}(10^{n-1} - a_{n-1}) + \cdots \\ &\qquad + a_1(10 - a_1) + a_0(1 - a_0) \end{aligned}$$

is positive. For, since n is at least 2, 10^n is at least 100, and the smallest value the first term can have is 99. All terms except possibly the last are non-negative, and the smallest value the last term can have is $9(-8) = -72$. Thus $N - N_1 = N - S(N) \geq 99 - 72 = 27$; that is, a number N with three or more digits is bigger, by at least 27, than the term in the sequence which follows it.

This means that the sequence

$$N,\ N_1,\ N_2,\ \cdots$$

decreases monotonically so long as the terms have at least three digits. But in such a decreasing sequence, eventually a term with less than three digits must turn up. It will have one or two digits, and part (ii) of our proof must then be used.

(ii) The writer knows of no way of showing that the assertion holds for $0 < N < 100$ except by direct enumeration. He suggests the reader work Exercise 6 to verify our claim.

V. If the number N occurs in Table 10.1 (constructed according to the rules given on p. 75) then $2N + 1$ is not a prime number; if N does not occur in the table, then $2N + 1$ is a prime number.

Proof: We begin by finding a formula for the entries in Table 10.1. The first number in the n-th row is

$$4 + (n - 1)3 = 3n + 1.$$

The common difference of the arithmetic progression comprising the n-th row is $2n + 1$; hence the m-th number of the n-th row is

$$3n + 1 + (m - 1)(2n + 1) = (2m + 1)n + m.$$

Now if N occurs in the table, then $N = (2m + 1)n + m$ for some pair of integers m and n. Therefore

$$2N + 1 = 2(2m + 1)n + 2m + 1 = (2m + 1)(2n + 1)$$

is composite.

Next we must show that, if N is not in Table 10.1, $2N + 1$ is prime; or, equivalently, if $2N + 1$ is not prime, N is in the table. So, suppose $2N + 1 = a \cdot b$, where a, b are integers greater than 1.

Since $2N + 1$ is odd a and b must both be odd, say

$$a = 2p + 1, \qquad b = 2q + 1,$$

so that

$$2N + 1 = ab = (2p + 1)(2q + 1) = 2p(2q + 1) + 2q + 1$$

and

$$N = (2q + 1)p + q.$$

But this means N appears as the q-th number of the p-th row in Table 10.1. We conclude that $2N + 1$ is a prime number if and only if N does *not* occur in Table 10.1.

EXERCISES

1. Prove identity (1), p. 72, inductively, using the following steps:

   (i) Show that

   $$\sum_{i=1}^{k} i = \tfrac{1}{2}k(k+1) \qquad \text{for every integer } k > 0.$$

   (ii) Using the result of (i), compute the difference

   $$\Big(\sum_{i=1}^{k+1} i\Big)^2 - \Big(\sum_{i=1}^{k} i\Big)^2$$

   and show it is equal to $(k+1)^3$.

   (iii) Conclude that identity (1) for k implies identity (1) for $k + 1$. [Just add $(k+1)^3$ to both sides of the assumed identity and use (ii) to obtain the desired identity.]

   (iv) Check that (1) holds for $k = 1$.

2. Show that there are 705 non-equivalent four-digit numbers with not all four digits the same (using the definition of equivalence on p. 79).

3. (a) Show that the difference $D_1 = 999(2) + 90(1)$ obtained for $a - d = 2$, $b - c = 1$, is equivalent to the difference $E_1 = 999(8) + 90(1)$ for $a - d = 8$, $b - c = 1$.

(b) Show that, if M_1 has $b - c = 1$, $a - d = n$ and M_2 has $b - c = 1$, $a - d = 10 - n$, then $T(M_1)$ and $T(M_2)$ are equivalent.

(c) Show that, if M_1 has $a - d = 9$, $b - c = k$ and M_2 has $a - d = 9$, $b - c = 10 - k$, then $T(M_1)$ and $T(M_2)$ are equivalent.

4. Write a computer program for the transformation T. Feed in a number K of Table 10.2 and instruct the machine to feed in successive numbers $T(K)$, $T^2(K)$, $\cdots$ until 6174 turns up, and to record the number of steps necessary to achieve this for each of the 30 entries in Table 10.2.

5. Starting with a cycle of eight arbitrary positive integers, verify Ducci's observation. In how many steps did you arrive at a cycle of eight equal numbers in your particular example?

6. (a) Complete four more rows of the "addition of squares" Table 10.3 by entering the number $i^2 + j^2$ in the i-th row and j-th column.

n		0	1	2	3	4	5	6	7	8	9
	n^2	0	1	4	9	16	25	36	49	64	81
0	0	0	1	4	9	16	25	36	49	64	81
1	1		2	5	10	17	26	37	50	65	82
2	4			8	13	20	29	40	53	68	85
3	9				18	25	34	45	58	•	•
•	•				•	•	•	•	•	•	•
•	•										
•	•										

Table 10.3

(b) Verify that 4, 16, 37, 58, 89, 145, 42, 20 is a repeating cycle $N_l, N_{l+1}, \cdots, N_{l+7}, N_{l+8} = N_l$.

(c) With the aid of Table 10.3, calculate

$$S(N) = S(10t + u) = u^2 + t^2 = N_1,$$

then $N_2, \cdots$, etc. for all $N < 100$ (i.e., $0 \leq u \leq 9$, $0 \leq t \leq 9$). Take advantage of labor saved by the fact that $u^2 + t^2 = t^2 + u^2$, and by the fact that you can stop calculating the moment you hit one of the numbers in the repeating cycle.

7. The columns in Table 10.4 contain the integers in consecutive order; each column is read downward, and we begin with the left-most column. For the row between bars, prove

(a) all the numbers in the row are of the form $n^2 - n + 1$;

(b) beginning with 3, every 3rd number is a multiple of 3;

(c) beginning with 7, every 7th number is a multiple of 7;

(d) beginning with 13, every 13th number is a multiple of 13;

(e) the product of any two consecutive numbers in the row is also in the row, and the position in the row where this product occurs is one more than the square in the bottom entry of the column of the smaller factor in the product; e.g., $7 \cdot 13 = 91$ is in the $(9 + 1)$-th position, where 9 is the square at the bottom of the column in which 7 occurs.

										·	·
									82	·	·
								65	83	·	·
							50	66	84	·	·
						37	51	67	85	·	·
					26	38	52	68	86	·	·
				17	27	39	53	69	87	·	·
			10	18	28	40	54	70	88	·	·
		5	11	19	29	41	55	71	89	·	·
	2	6	12	20	30	42	56	72	90	·	·
1	3	7	13	21	31	43	57	73	91	·	·
	4	8	14	22	32	44	58	74	92	·	·
		9	15	23	33	45	59	75	93	·	·
			16	24	34	46	60	76	94	·	·
				25	35	47	61	77	95	·	·
					36	48	62	78	96	·	·
						49	63	79	97	·	·
							64	80	98	·	·
								81	99	·	·
									100	·	·
										·	·

Table 10.4

8. Each column of Table 10.5 is an arithmetic progression. The common differences are, respectively, the odd integers, 1, 3, 5, $\cdots$. The first term of the n-th column is the sum of the first n terms of the arithmetic progression

$$1, 5, 9, 13, 17, \cdots,$$

that is, the n-th partial sum of the series

$$1+5+9+13+17+\cdots.$$

Prove that the sum of the numbers in the k-th *row* is k^3.

1								
2	6							
3	9	15						
4	12	20	28					
5	15	25	35	45				
6	18	30	42	54	66			
7	21	35	49	63	77	91		
8	24	40	56	72	88	104	120	
•	•	•	•	•	•	•	•	•
•	•	•	•	•	•	•	•	• •

Table 10.5

REFERENCE

H. Steinhaus, 100 *Problems in Elementary Mathematics*, Basic Books, 1964, New York.

ESSAY ELEVEN

A Problem of Regiomontanus

Suppose a statue, h feet high, stands on a pedestal p feet high (see Figure 11.1). A man, whose eye-level is e feet above ground, walks toward the statue, gazing at it. At what distance from the base of the statue should he stand to make the statue appear the largest (i.e., so that his lines of sight to the bottom and top of the statue subtend the largest angle)?†

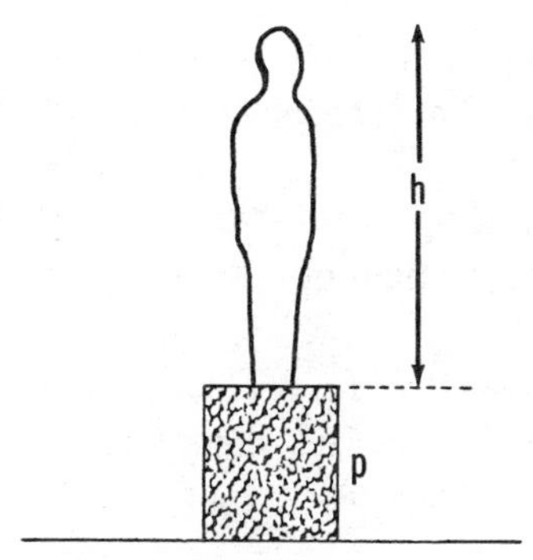

Figure 11.1

† A problem of this kind was first posed in 1471 by the mathematician Johannes Müller, known as Regiomontanus. The solution given here (which is simpler than the solution by calculus) is due to A. Lorsch.

The problem is interesting only if the beholder's eye-level is below the top of the pedestal, or above the statue. We assume from now on that $e < p$. (The case $e > p + h$ could be treated similarly; and for $p < e < p + h$, see Exercise 1.)

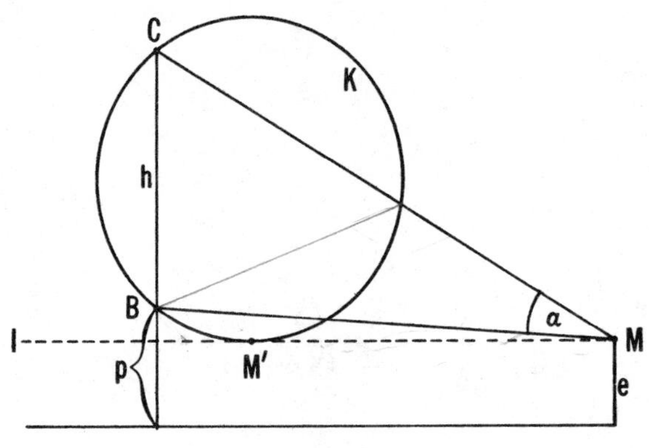

Figure 11.2

Figure 11.2 is a schematic diagram of our situation. We have dropped a perpendicular from the highest point C of the statue to the ground; we have denoted by B the bottom of the statue, by l the locus of the beholder's eyes as he walks along the ground, by M an arbitrary position of the beholder, and by α the angle BMC which is to be maximized as M moves along l. Furthermore, we construct the circle K through the points B and C and tangent to the line l.

Claim: Among all points M on l, the one for which $\alpha = \measuredangle BMC$ is largest is the point M' of tangency of the circle K with the line l.

Proof: In order to show that $\measuredangle BM'C > \measuredangle BMC$ for all points $M \neq M'$ on line l (to the right of the statue), we shall use the following facts:

Let β denote the angle $BM'C$. Then for any point P on the arc $BM'C$ we have $\measuredangle BPC = \beta$; see Figure 11.3. It is not difficult to prove (see Exercise 2) that

(a) if R is any point in the interior of the circle, then $\measuredangle BRC > \beta$,

(b) if Q is any point in the exterior of the circle (and on the same side of the line through B and C as the major arc), then $\measuredangle BQC < \beta$.

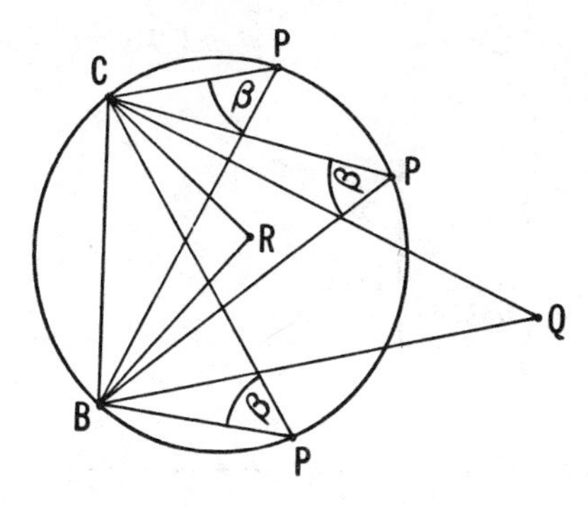

Figure 11.3

Since line l has only one point, namely M', in common with circle K, and since all other points of l right of the statue are outside K, it follows from statement (b) above that the angle BMC is indeed maximized for $M = M'$.

EXERCISES

1. Suppose the eye-level e of the beholder is greater than the height p of the pedestal, but less than $p + h$. Show that then the angle α of the lines of sight is a maximum when the beholder gets as close as he can to the statue.

2. Prove statements (a) and (b) above. [*Hint*: Connect the point where the line through C and R (C and Q) intersects the circle with the point B.]

3. Let l be the x-axis of a Cartesian coordinate system, let the line through B and C be the y-axis. Set $b = p - e$, $c = p + h - e$, so that the coordinates of B and C are $(0, b)$ and $(0, c)$, respectively see Figure 11.4.

   (i) Prove that the center of the circle K has coordinates

   $$[\sqrt{bc},\ (b + c)/2] \text{ and radius } (b + c)/2.$$

   (ii) Show that $\sqrt{bc} < (b + c)/2$, by interpreting each as a length in Figure 11.4.

   (iii) Let b and c be any two positive numbers. Show algebraically that

$$(*) \qquad \sqrt{bc} \leq \frac{b + c}{2}$$

and that equality holds if and only if $b = c$. [*Hint*: Observe that $(b - c)^2$ is always non-negative.]

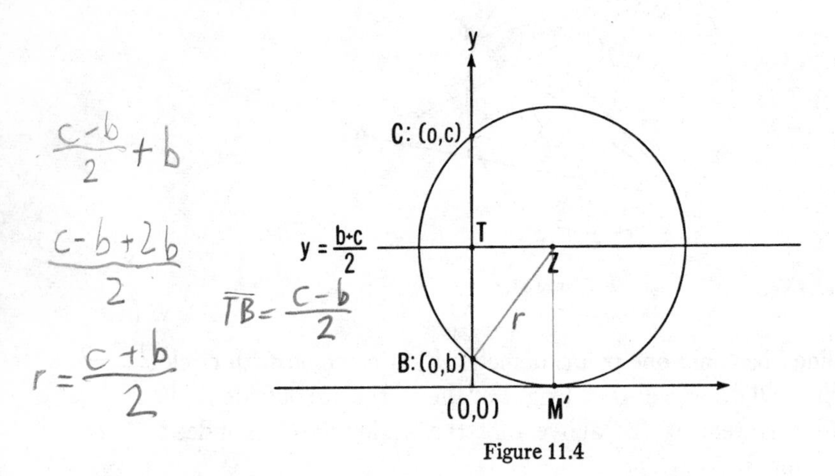

Figure 11.4

Remark: The inequality (*) is the important *inequality between the geometric and arithmetic means* of two positive numbers. The solution of part (iii) of Exercise 3 shows the general validity of inequality (*), and the solution of part (ii) gives a geometric illustration of this famous inequality.

4. If a 10′ statue stands on a 13′ pedestal, where should a man whose eye-level is 5′ above ground stand in order that the statue appear to him as large as possible?

REFERENCE

H. Dorrie, 100 *Great Problems of Elementary Mathematics*, Dover, 1958, New York.

ESSAY TWELVE

Complementary Sequences

I. Beatty's Theorem. In 1926, Sam Beatty of the University of Toronto (one of my teachers) made a remarkable discovery concerning sequences of irrational numbers. Suppose X is a positive irrational number, say $\sqrt{2}$, for example. The reciprocal of X we call Y; in this case $Y = 1/\sqrt{2} = \sqrt{2}/2$, and is approximately .7. Adding 1 to each of X and Y we obtain

$$1 + X \approx 2.4 \qquad \text{and} \qquad 1 + Y \approx 1.7.$$

(The symbol $\approx$ means "is approximately equal to".) Now a table of the approximate multiples of $1 + X$ and $1 + Y$ is made, and these multiples are plotted on a number axis (see Table 12.1 and Figure 12.1). Then we see that in each interval $(n, n + 1)$ between a pair of consecutive positive integers there occurs exactly one of the numbers from the table.

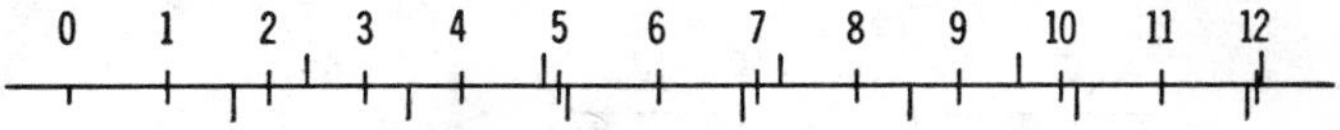

0 1 2 3 4 5 6 7 8 9 10 11 12

Figure 12.1

n	1	2	3	4	5	6	7	$\cdots$
$n(1+X)$	2.4	4.8	7.2	9.6	12^+	14.4	16.8	$\cdots$
$n(1+Y)$	1.7	3.4	5.1	6.8	8.5	10.2	11.9	$\cdots$

Table 12.1

In general, we have the theorem:

THEOREM 1. *Let X be any positive irrational number and Y its reciprocal. Then the two sequences*

$$\begin{aligned}&1+X,\quad 2(1+X),\quad 3(1+X),\quad \cdots,\\ &1+Y,\quad 2(1+Y),\quad 3(1+Y),\quad \cdots\end{aligned}$$

together contain exactly one number from each of the intervals $(n, n+1)$ *between consecutive positive integers* $(n = 1, 2, 3, \cdots)$.

A beautiful proof of this theorem was published in 1927 by A. Ostrowski (Basel, Switzerland) and A. C. Aitken (Glasgow).

Proof: First we note that, since X and Y are irrational, so is every term of the sequences in question. This means in particular that no term is an integer.

Now it is very easy to count the number of multiples of $1 + X$ which lie below a given positive integer N. We simply divide N by $1 + X$ and drop the fractional part of the result. That is to say, the number of multiples of $1 + X$ which are less than N is $[N/(1 + X)]$, where the symbol $[z]$ denotes the greatest integer not exceeding z.

Similarly the number of terms in the second sequence (multiples of $1 + Y$) lying between 1 and N is $[N/(1 + Y)]$. Altogether then there are $[N/(1 + X)] + [N/(1 + Y)]$ terms of our sequences which lie between 1 and N.

Since $N/(1 + X)$ and $N/(1 + Y)$ are not integers, we have

$$\frac{N}{1+X} - 1 < \left[\frac{N}{1+X}\right] < \frac{N}{1+X}$$

and

$$\frac{N}{1+Y} - 1 < \left[\frac{N}{1+Y}\right] < \frac{N}{1+Y}.$$

Adding these inequalities, and observing that

$$\frac{1}{1+X} + \frac{1}{1+Y} = \frac{1}{1+X} + \frac{1}{1+\frac{1}{X}} = \frac{1}{1+X} + \frac{X}{1+X} = 1,$$

we obtain

$$N - 2 < \left[\frac{N}{1+X}\right] + \left[\frac{N}{1+Y}\right] < N.$$

Since

$$\left[\frac{N}{1+X}\right] + \left[\frac{N}{1+Y}\right]$$

is an integer, it follows that

$$\left[\frac{N}{1+X}\right] + \left[\frac{N}{1+Y}\right] = N - 1.$$

This means that the total number of terms of our sequences which lie below the positive integer N is $N - 1$.

Similarly, the total number of terms up to $N + 1$ is N. That is to say, if we increase the integer N by 1, another term of one of the sequences is admitted, implying that exactly one term lies between N and $N + 1$ (QED).

It follows from the theorem that if we drop the fractional parts of the terms of our sequences, each term yields an integer and each natural number occurs exactly once. Thus we have the

COROLLARY. *The sequences* $[n(1 + X)]$, $[n(1 + Y)]$, *called Beatty sequences corresponding to the irrational number* X, *together contain each natural number exactly once.*

Pairs of sequences having this property are called *complementary.*

II. Lambek and Moser. In the summer of 1954 two Canadian mathematicians, J. Lambek (McGill) and Leo Moser (Alberta), took up this subject of complementary sequences and made some interesting discoveries.

Let us suppose that we have some non-decreasing sequence of non-negative integers, such as the famous Fibonacci sequence, for example. We denote it by $f(n)$. From this given sequence $f(n)$ we can construct another sequence $f^*(n)$ according to the definition

$$f^*(n) = \text{the number of positive integers } m \text{ such that } f(m) < n$$

(see Table 12.2); that is, the new sequence f^* is simply the "frequency distribution function" of f in the sense of the number of values m for which f takes values less than n.

n	1	2	3	4	5	6	7	8	9	10	11	12	13	14	...
$f(n)$	1	1	2	3	5	8	13	21	34	55	89	144	233	377	...
$f^*(n)$	0	2	3	4	4	5	5	5	6	6	6	6	6	7	...
$F(n)$	2	3	5	7	10	14	20	29	43	65	100	149	246	391	...
$G(n)$	1	4	6	8	9	11	12	13	15	16	17	18	19	21	...

Table 12.2.

Clearly there are no terms of $f(n)$ less than 1, 2 of them are less than 2, 4 of them less than 5, 5 less than 6, 5 less than 7, 5 less than 8, and so on. Obviously f^* is nondecreasing.

We observe a surprising result: if one begins with f^* and works out its frequency distribution, the original sequence f is obtained; that is, $f^{**} = f$. (Check it.) This accounts for the terminology that f and f^* are *inverse* sequences.† The condition for f and f^* to be inverse is simply that f be non-decreasing. A precise formulation of this and the results stated below is given in the appendix at the end of this essay. A detailed proof, suggested by H. Shapiro, is included there.

Now make up two new sequences, $F(n)$ and $G(n)$, from f and f^* as follows: simply add n to each; that is, for each n, form

$$F(n) = f(n) + n \qquad \text{and} \qquad G(n) = f^*(n) + n.$$

Then F and G are complementary sequences. (See definition on p. 95.)

Moreover, the converse is also true: If we separate the natural numbers into two increasing sequences F and G, then the sequences formed by subtracting n, that is, $f(n) = F(n) - n$ and $f^*(n) = G(n) - n$, are inverse. (Lambek and Moser found out much more than this.)

We relegate the proofs of these results to the appendix and apply them to the following intriguing problem:

† See Exercise 1.

III. The Non-Squares. If we separate the natural numbers into the squares and the non-squares, we obtain two complementary sequences

$$F(n): \quad 1,\ 4,\ 9,\ 16,\ 25,\ 36,\ 49,\ 64,\ \cdots;$$

$$G(n): \quad 2,\ 3,\ 5,\ 6,\ 7,\ 8,\ 10,\ 11,\ 12,\ 13,\ 14,\ 15,\ 17,\ \cdots.$$

The question is, what is the general term of the sequence $G(n)$; that is, *find a formula for the n-th non-square.*

The first thing to do is to form the inverse sequences f and f^*, as defined above. Once we have a formula for $f^*(n)$, the formula

$$G(n) = f^*(n) + n$$

for the n-th non-square is immediately obtained. To this end, we recall the definition

$$f^*(n) = \text{the } \textit{number} \text{ of positive } m \text{ such that } f(m) < n.$$

We also have that

$$f(m) = F(m) - m,$$

where $F(m) = m^2$ is the sequence of perfect squares. Consequently,

$$f^*(n) = \text{the } \textit{number} \text{ of positive } m \text{ such that } m^2 - m < n.$$

The sequence of integers $m^2 - m = m(m-1)$ is increasing for $m = 1, 2, \cdots$ (it is a product of increasing factors), so the number of integers m such that $m^2 - m < n$ is the *biggest* integer m satisfying $m^2 - m < n$. Thus

$$f^*(n) = \text{the biggest integer } m \text{ such that } m^2 - m < n.$$

Because m and n are integers, the greatest m satisfying $m^2 - m < n$ is also the greatest m which satisfies $m^2 - m < n - \frac{1}{4}$. (In fact, both inequalities are equivalent to $m^2 - m \leq n - 1$.) Consequently, the number we seek is the greatest solution of

$$m^2 - m + \tfrac{1}{4} < n,$$

that is, of

$$(m - \tfrac{1}{2})^2 < n,$$

and therefore of

$$m - \tfrac{1}{2} < \sqrt{n}$$

and of

$$m < \sqrt{n} + \tfrac{1}{2}.$$

But the greatest m less than $\sqrt{n} + \frac{1}{2}$ is by definition $[\sqrt{n} + \frac{1}{2}]$, the "greatest integer" function we encountered in the proof of Beatty's Theorem. Thus a formula for the n-th non-square is

$$G(n) = f^*(n) + n = [\tfrac{1}{2} + \sqrt{n}] + n.$$

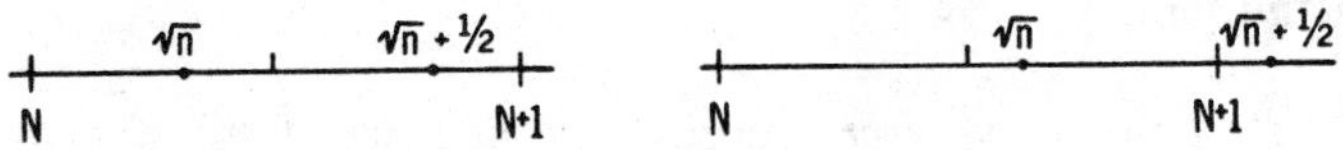

Figure 12.2

We observe that if $\sqrt{n}$ lies in the first half of an interval between positive integers N and $N + 1$, then $[\frac{1}{2} + \sqrt{n}] = N$; and if it lies in the second half, then $[\frac{1}{2} + \sqrt{n}] = N + 1$ (see Figure 12.2). In any event, then, we have that the value of $[\frac{1}{2} + \sqrt{n}]$ is the integer *closest* to $\sqrt{n}$, which we denote by $\langle\sqrt{n}\rangle$. Consequently, the formula for the n-th non-square is given by

$$G(n) = n + \langle\sqrt{n}\rangle.$$

IV. A Remarkable Equality. A slightly different approach to this problem leads to a different formula for the n-th non-square and provides a most unexpected relation.

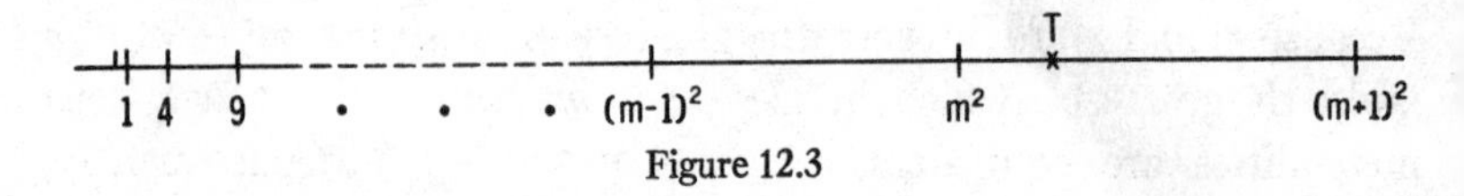

Figure 12.3

Suppose T is the n-th non-square, and that the squares nearest to it are m^2 and $(m + 1)^2$ (see Figure 12.3). Thus the square root of T

lies between m and $m+1$. This means that

$$[\sqrt{T}] = m. \tag{1}$$

Now of the T positive integers $1, 2, \cdots, T$, m are perfect squares. This means that $T-m$ of them are non-squares; that is, T is the $(T-m)$-th non-square. Since it is the n-th non-square also, we get

$$T - m = n,$$

so that

$$T = m + n.$$

Thus $m = [\sqrt{T}]$ becomes

$$m = [\sqrt{m+n}]$$

and we have

$$\begin{aligned} T &= n + m \\ &= n + [\sqrt{n+m}]. \end{aligned}$$

Of course, the m in this is also equal to $[\sqrt{n+m}]$. We may carry on, then, with

$$\begin{aligned} T &= n + [\sqrt{n + [\sqrt{n+m}]}], \\ &= n + [\sqrt{n + [\sqrt{n + [\sqrt{n+m}]}]}] \\ &= \cdots \end{aligned} \tag{2}$$

as far as we please to go. (Lambek and Moser dealt brilliantly with this problem of limits.)

What makes this useful is that, beyond the second stage, the m has no effect at all; that is, the expression

$$T = n + [\sqrt{n + [\sqrt{n}]}]$$

gives the same result for T as every expression in (2). This, then, is the value of T. Accepting this momentarily, we obtain, by comparison with the formula $n + \langle\sqrt{n}\rangle$ derived in subsection III, the remarkable equation referred to in the heading,

$$[\sqrt{n + [\sqrt{n}]}] = \langle\sqrt{n}\rangle, \tag{3}$$

i.e., the expression on the left always yields the integer closest to $\sqrt{n}$.

We now show that the number

$$Z = n + [\sqrt{n + [\sqrt{n}]}]$$

is the same as any of the expressions (2) for T. Beginning with $n = T - m$, we observe that n is less than T. On the other hand, n cannot be as small as $(m - 1)^2$. To see this, we recall that $m^2 < T$ (see Figure 12.3), and from $m - 1 < m$, we deduce that

$$(m - 1)^2 \leq m(m - 1) = m^2 - m \qquad \text{for } m \geq 1.$$

Thus

$$(m - 1)^2 \leq m^2 - m < T - m = n.$$

Consequently, the value of n lies somewhere between $(m - 1)^2$ and T (see Figure 12.3).

The number m^2 also occurs in this interval and divides it into two parts. We consider these separately.

(i) Suppose $m^2 \leq n < T$; then

$$m \leq \sqrt{n} \qquad \text{and} \qquad [\sqrt{n}] = m.$$

In this case

$$\begin{aligned} Z &= n + [\sqrt{n + [\sqrt{n}]}] \\ &= n + [\sqrt{n + m}] = T. \end{aligned}$$

(ii) If $(m - 1)^2 < n < m^2$, then

$$\sqrt{n} > m - 1 \qquad \text{and} \qquad [\sqrt{n}] = m - 1.$$

This time

$$Z = n + [\sqrt{n + m - 1}] = n + [\sqrt{T - 1}].$$

Since $m^2 < T < (m + 1)^2$, we have that $m^2 \leq T - 1 < (m + 1)^2$ and $m \leq \sqrt{T - 1} < m + 1$ so that $[\sqrt{T - 1}] = m$. Thus $Z = n + m = T$.

V. Wythoff's Game. In 1907 the mathematician Wythoff invented a game for two players in which they alternately remove matches from two piles. Initially each pile contains an arbitrary number of matches, say m and n. We can keep track of the state of affairs with ordered pairs (a, b) denoting the numbers of matches remaining in each pile after a move has been made. The game begins, then, with (m, n).

The rules of the game dictate that a move must be one of three types:

(i) a withdrawal from pile I,

(ii) a withdrawal from pile II,

(iii) a withdrawal from both piles, the *same* number of matches from each.

These rules are expressed algebraically as follows: Change the pair (m, n) to one of the pairs

$$\text{(i)}\ (m-t, n), \qquad \text{(ii)}\ (m, n-t), \qquad \text{(iii)}\ (m-t, n-t);$$

in all cases $t \geq 1$, for at least one match must be withdrawn. The number of matches withdrawn is up to the player; he may take the whole pile if he likes. The winner is the person who takes the last match.

A sample game may look like this: After A has played

A leaves $(16, 13)$;	B plays to $(12, 9)$;
A plays to $(5, 9)$;	B plays to $(2, 6)$;
A plays to $(2, 1)$;	B must play to one of $(1, 1)$, $(0, 1)$, $(2, 0)$ or $(1, 0)$.
A Plays to $(0, 0)$ and wins.	

Notice that the pair $(2, 1)$ achieved by A on his next to the last turn is a winning position because, no matter which of the four possible moves B makes, A can win on his next move.

In general we define a pair (a, b) to be a *winning position* for player A if there is a strategy such that, no matter how B moves out of the position (a, b), A can win the game (not necessarily in one move). For example, $(2, 1)$ in the game above is a winning position. All four permissible moves from $(2, 1)$, lead to *losing positions* for B, i.e., to positions from which his opponent can win.

Beginning with the ultimate winning position $(0,\ 0)$, one can construct a list of all winning positions in Wythoff's game, for example, by examining all pairs $(x,\ y)$ for $x + y = 1,\ 2,\ \cdots$. In this way it can be shown that all pairs $(x,\ y)$ are either winning positions or losing positions (see Exercise 5). However, it turns out that this cumbersome procedure can be completely by-passed. Wythoff himself characterized all winning positions right off the bat by a formula. In the remainder of this essay, we shall give the characterization of the set W of all winning pairs, and we shall prove that the members of W are indeed winning positions by showing that every move from a position in W leads to a position *not* in W, and that there exists a move which brings a position not in W into one in W.

THEOREM 2. *The winning positions are given by the pairs* $(0,\ 0)$ *and* $(a_n,\ b_n)$, $n = 1,\ 2,\ \cdots$, *where* $\{a_n\}$ *and* $\{b_n\}$ *are the Beatty sequences corresponding to the irrational number*

$$X = \frac{-1 + \sqrt{5}}{2}. \tag{4}$$

We observe that X is the famous *golden mean*† and that

$$Y = \frac{1}{X} = \frac{2}{-1 + \sqrt{5}} = \frac{1 + \sqrt{5}}{2} = 1 + X.$$

We set

$$\tau = 1 + X = \frac{1 + \sqrt{5}}{2} = Y,$$

find that

$$1 + Y = \frac{3 + \sqrt{5}}{2} = \tau^2 = 1 + \tau,$$

† The golden mean is the ratio of the shorter to the longer side of a golden rectangle. Equivalently, it is the positive solution of the equation

$$\frac{1 - x}{x} = \frac{x}{1},$$

that is, of $x^2 + x - 1 = 0$.

and that the corresponding Beatty sequences $[n(1+X)]$, $[n(1+Y)]$ are

$$(5) \qquad a_n = [n\tau], \qquad b_n = [n\tau^2], \qquad n = 1, 2, \cdots.$$

If (x, y) is a pair in W, then so is (y, x) since the order we assign to our two piles of matches does not matter. It will be convenient to denote the smaller number of a pair by a_n and to write the pair (a_n, b_n), with a_n first.

Proof of Theorem 2: First we show that the pairs

$$(6) \qquad (a_n, b_n) = ([n\tau], [n\tau^2])$$

have the properties

(i) $b_n - a_n = n$ for $n = 1, 2, \cdots$, $(a_0, b_0) = (0, 0)$,

(ii) the smaller member, a_n, of the n-th pair is the smallest integer not used in any previous pair;

and conversely, pairs satisfying properties (i) and (ii) are of the form (6).

To establish property (i) we write

$$b_n - a_n = [n\tau^2] - [n\tau].$$

Recalling that $\tau^2 = 1 + \tau$, we have

$$b_n - a_n = [n(1+\tau)] - [n\tau] = [n + n\tau] - [n\tau],$$

and since n is a positive integer, $[n + n\tau] = n + [n\tau]$, so

$$b_n - a_n = n + [n\tau] - [n\tau] = n.$$

To establish property (ii), we recall that Beatty sequences are complementary (see the Corollary of Theorem 1 on page 95). This means that $a_i \neq b_j$ for all positive integers i, j, and that every positive integer belongs to one of the sets $\{a_i\}$ or $\{b_i\}$. Now let c be the smallest positive integer *not* in $\{a_k\}$, *nor* in $\{b_k\}$, for $k = 1, 2, \cdots, n-1$. Then it must be either in $\{a_k\}$ or in $\{b_k\}$ for $k \geq n$. But the smallest integer in these sets is a_n because both sequences are strictly increasing and $a_j < b_j$ for all $j > 0$. Thus $c = a_n$.

To prove the converse, we merely observe that properties (i) and (ii) can be used as a recipe for building pairs recursively. The n-th pair is $(p, \; p+n)$ where p is the smallest integer not used previously. Since this recipe leads to a unique determination of pairs, we may conclude that the Beatty sequences are the only sequences yielding pairs having properties (i) and (ii).

Next we show that the set W of pairs $(a_n, \; b_n)$ of the form (6) indeed constitutes all winning positions in the sense that every move from such a pair leads to a pair not in W, and given a pair $(c, \; d)$ not in W there is a move which leads to a pair in W.

Let (a_n, b_n) be a given pair in W. After a permissible move, one of the following pairs may result:

$$1) \; (a_n - t, \; b_n), \qquad 2) \; (a_n, \; b_n - t), \qquad 3) \; (a_n - t, \; b_n - t).$$

Pairs of the form 1) and 2) are not in W because each contains a number belonging to the n-th pair, and every positive integer occurs in one and only one pair of W; e.g., if b_n occurs in $(a_n, \; b_n)$ it cannot occur in any other winning position. A pair of the form 3) is not in W because only the n-th pair $(a_n, \; b_n)$ of W [after $(0, \; 0)$] has members differing by n. But $(a_n - t, \; b_n - t)$ is not the n-th pair $(t \neq 0)$, yet its members differ by n. We conclude that all positions obtainable from a position in W by a permissible move are outside the set W.

Next, suppose a pair $(c, \; d)$ is not in W. We shall show that we can determine a number s such that at least one of the pairs

$$1) \; (c - s, \; d) \qquad 2) \; (c, \; d - s), \qquad 3) \; (c - s, \; d - s)$$

is in W.

In case $c = d$, choose $s = c = d$ and arrive at the ultimate winning position $(0, \; 0)$ in one move: $(c - s, \; d - s) = (0, \; 0)$.

Suppose $c \neq d$; then name the numbers so that $c < d$. Now every positive integer is in exactly one of the Beatty sequences, so for the integer c, either

$$\text{I: } c = a_k \text{ for some } k, \qquad \text{or} \qquad \text{II: } c = b_l \text{ for some } l;$$

i.e., c belongs to one of the winning positions

$$\text{I: } (c, \; b_k) = (a_k, \; b_k) \qquad \text{or} \qquad \text{II: } (a_l, \; c) = (a_l, \; b_l).$$

Case I. If $b_k < d$, set $s = d - b_k$ and move to

$$(c,\ d - s) = (a_k,\ b_k)$$

in W.

If $d < b_k$, then $c < d < b_k$ implies $0 < d - c < b_k - c = k$. Calculate the positive integer $n = d - c$; by the last inequality, $n < k$. Set

$$\begin{aligned} s &= c - a_n \ (>0 \text{ since } a_k > a_n) \\ &= c - (b_n - n) \ (\text{Property (i)}) \\ &= c - b_n + (d - c) \ (\text{definition of } n) \\ &= d - b_n. \end{aligned}$$

Move to $(c - s,\ d - s) = (a_n, b_n)$ in W.

Case II. In this case it follows that $a_l < d$ because $a_l < b_l$ for all l, and $b_l = c < d$ by hypothesis. Set $s = d - a_l$, and move to $(c,\ d - s) = (b_l,\ a_l)$ in W.

We conclude that there is always a permissible move which brings a non-winning pair into a winning pair.

EXERCISES

1. Let $f(x)$ be a strictly increasing function for $x \geq 0$. Let

$$f^*(y) = \max \{x \geq 0 \mid f(x) \leq y\}.$$

Show that f and f^* are inverse functions, i.e., that $f^*(f(x)) = x$.

2. Show that the n-th non-triangular number is $n + \langle \sqrt{2n} \rangle$. N is a *triangular number* if it is the sum of consecutive integers beginning with 1.

3. Give a direct proof of the identity

$$[\sqrt{n + [\sqrt{n}]}] = \langle \sqrt{n} \rangle.$$

[*Hint*: Treat the cases (a) $\sqrt{n}$ is in the first half, (b) $\sqrt{n}$ is in the second half of an interval between successive integers separately; i.e., (a) $k \leq \sqrt{n} < k + \frac{1}{2}$, and (b) $k + \frac{1}{2} \leq \sqrt{n} < k + 1$.]

4. Show that the n-th positive integer *not* of the form $[e^m]$, $m \geq 1$, is $T = n + [\log (n + 1 + [\log (n + 1)])]$. Here e denotes the base of natural logarithms, which appear in the formula for T.

5. Prove that every pair of integers (x, y) represents either a winning or a losing position in Wythoff's game. [*Hint*: Examine all positions involving zero, then one, then two matches. Then prove, by mathematical induction on the number of matches, that if the assertion is true for all pairs (x, y) with $x + y < n$, then it follows for all pairs (x, y) with $x + y = n$.]

REFERENCES

S. Beatty, Problem 3173, American Mathematical Monthly, vol. 33, 1926, p. 159.

H. S. M. Coxeter, Scripta Mathematica, vol. 19, 1953, p. 135.

H. W. Gould, "Non-Fibonacci Numbers", Fibonacci Quarterly, vol. 3, p. 177.

Lambek and Moser, American Mathematical Monthly, vol. 61, 1954, p. 454.

T. O'Beirne, *Puzzles and Paradoxes*, Oxford, p. 130–138.

Ostrowski and Aitken, Solution to Problem 3173, American Mathematical Monthly, vol. 34, p. 159.

Appendix

THEOREM. Let $f(n)$ be a non-decreasing function which maps the set of non-negative integers into itself. Let

$$f^*(n) = \begin{cases} \text{the number of positive solutions } k \text{ (i.e., } k \geq 1) \\ \text{of the inequality } f(k) < n. \end{cases}$$

Then

(a) The functions

$$F(n) = f(n) + n, \qquad G(n) = f^*(n) + n$$

are *complementary*; that is, if R_F, R_G denote the ranges (set of values) of F and G, respectively, for $n = 1, 2, 3, \cdots$, then

(i) every integer ≥ 1 is either in R_F or in R_G,
(ii) no integer ≥ 1 is in both R_F and R_G.

(b) Conversely, if $F(n)$, $G(n)$ are increasing functions satisfying conditions (i) and (ii), then the functions

$$f(n) = F(n) - n, \qquad g(n) = G(n) - n$$

have the properties

$$g(n) = f^*(n) \qquad \text{and} \qquad g^*(n) = f(n)$$

for $n \geq 1$.

(c) The functions $f(n)$ and $f^*(n)$ are *inverse*, that is, they satisfy

$$\text{(iii)}\ f^{**}(n) = f(n) \qquad \text{for } n = 1,\ 2,\ \cdots.$$

Proof of (a): It will be helpful, in what follows, to use the following equivalent definition of $f^*(n)$:

$$f^*(n) = \begin{cases} \text{the largest integer } k \text{ such that } f(k) < n, \\ \quad \text{if such an integer exists;} \\ 0 \text{ if no such integer } k \text{ exists.} \end{cases}$$

Since, by hypothesis, $f(n)$ is non-decreasing and, by definition, $f^*(n)$ is non-decreasing, it follows that $F(n)$ and $G(n)$ are strictly increasing functions.

Suppose $m \geq 1$ is not in R_F; i.e., there is no s such that $m = f(s) + s$. Then either

$$m \leq f(0) + 0 = f(0),$$

in which case $f^*(m) = 0$, and $m = f^*(m) + m$ is in R_G; or

$$m > f(0) + 0,$$

in which case there is an integer l such that

$$f(l) + l < m < f(l + 1) + l + 1.$$

Since only integers are involved, these inequalities may be written

$$f(l) + l < m \leq f(l+1) + l$$

so that

$$f(l) < m - l \leq f(l+1). \tag{1}$$

In case $l = 0$, (1) yields $f(0) < m \leq f(1)$, so $f^*(m) = 0$ and again, $m = f^*(m) + m$ is in R_G. In case $l \geq 1$, (1) yields

$$l = f^*(m - l) \tag{2}$$

and since $m - l > f(l) \geq 0$, $m - l \geq 1$. It follows that

$$m = (m - l) + l = m - l + f^*(m - l)$$

is in R_G. This establishes condition (i).

Next we must show: if an integer $m \geq 1$ is in R_G, then it is not in R_F. Accordingly, let

$$m = f^*(r) + r, \qquad m \geq 1, \quad r \geq 1;$$

then

$$m - r = f^*(r). \tag{3}$$

If $m = r$, (3) yields $0 = f^*(r) = f^*(m)$, so there is no integer $k \geq 1$ for which $f(k) < m$. It follows that $f(1) \geq m$ and, since f is non-decreasing,

$$f(s) + s > m \qquad \text{for all } s \geq 1,$$

so m is not in R_F.

If $m - r \geq 1$, (3) yields (see definition of f^*)

$$f(m - r) < r \leq f(m - r + 1). \tag{4}$$

Adding $m - r$ to all members of (4) we obtain

$$m - r + f(m - r) < m \leq m - r + f(m - r + 1)$$
$$< m - r + 1 + f(m - r + 1)$$

and conclude, from

$$k + f(k) < m < k + 1 + f(k + 1) \qquad \text{for all } k = m - r \geq 1$$

that m is not in R_F. This proves condition (ii).

Proof of (*b*): To prove the converse, assume that the functions $A(n)$, $B(n)$ are increasing and complementary; i.e., they satisfy conditions (i) and (ii).

We set

$$a(n) = A(n) - n, \qquad b(n) = B(n) - n$$

and claim that

$$a^*(n) = b(n) \qquad \text{and} \qquad b^*(n) = a(n).$$

First we introduce the auxiliary function

$$C(n) = a^*(n) + n$$

and observe that the increasing character of A, B guarantee the non-decreasing character of a, a^* and b.

By hypothesis, the ranges R_A and R_B of A and B, respectively, for $n \geq 1$, are disjoint and cover the set of all positive integers. By part (a) of the theorem, $A(n) = a(n) + n$ and $C(n) = a^*(n) + n$ have disjoint ranges R_A and R_C which cover the set of all positive integers. It follows that

$$R_B \equiv R_C. \tag{5}$$

Now if two increasing functions $B(n)$ and $C(n)$ have the property that $B(1), B(2), \cdots$ and $C(1), C(2), \cdots$ cover the same set of integers [and this is what (5) says], then

$$B(n) = C(n) \qquad \text{for } n = 1, 2, \cdots;$$

that is,

$$b(n) + n = a^*(n) + n \qquad \text{for } n = 1,\ 2,\ \cdots,$$

and hence

$$b(n) = a^*(n) \qquad \text{for } n = 1,\ 2,\ \cdots.$$

If we reverse the roles played by $A(n)$ and $B(n)$, the same reasoning yields that

$$b^*(n) = a(n) \qquad \text{for } n \geq 1.$$

This establishes part (b) of our theorem.

Part (c) is now an easily proved corollary. Let $f(n)$ be a non-decreasing function and $f^*(n)$ as defined before. Then the functions

$$F(n) = f(n) + n, \qquad G(n) = f^*(n) + n$$

are complementary by part (a) of our theorem. They therefore satisfy the hypothesis of part (b) and hence the conclusion, which implies that

$$f(n) = f^{**}(n) \qquad \text{for } n \geq 1$$

as was to be demonstrated.

Part (c) can also be proved directly.

ESSAY THIRTEEN

Pythagorean Arithmetic

Pythagoras, a native of the Greek island of Samos (he lived from about 570 B.C. to 500 B.C.), migrated to Crotona in southern Italy where he founded an academy of learning which brought him a devoted following and lasting fame. All the discoveries of the school were, by custom, attributed to Pythagoras himself. Just what his personal contributions are is almost impossible to estimate. However, in total, the school's achievement was great, practically marking the advent of deductive mathematics. In the hands of the Pythagoreans mathematics was directed along various channels, some of which have not yet dried up. In this section, we consider several Pythagorean topics and some later developments.

In the terminology of the Greeks, "arithmetic" is equivalent to our number theory, while "logistic" was their term for practical calculations.

Amicable Numbers. Two positive integers constitute an *amicable pair* (friendly pair) if the proper divisors of each one add up to the other. (The proper divisors do not include the number itself.) The smallest pair, the only one known to the Pythagoreans, is (220, 284):

220 has divisors 1, 2, 4, 5, 10, 11, 20, 22, 44, 55, 110, whose sum is 284;
284 has divisors 1, 2, 4, 71, 142, whose sum is 220.

The next new pair was announced in 1636 by the celebrated French genius Pierre de Fermat (1601–1665); it is (17296, 18416). In 1638 Descartes gave a third pair. In 1747 Euler gave 30 pairs, and in 1750 he increased that number to 60 pairs. Today over 900 pairs are known. Surprisingly, the second smallest pair (1184, 1210) was overlooked by Fermat et al. It was discovered by a 16-year old Italian boy, Niccolò Paganini (not the violinist) in the second half of the nineteenth century.

There are various rules for finding amicable pairs. An Arabian mathematician stated that, for $x > 1$,

$$a = 3\cdot 2^x - 1,$$
$$b = 3\cdot 2^{x-1} - 1,$$
$$c = 9\cdot 2^{2x-1} - 1$$

yield the amicable pair $(2^x ab,\ 2^x c)$ provided that a, b, c are all prime numbers. For $x = 2$, this method yields (220, 284).

More recently amicable pairs have been generalized to *amicable chains*: the sum of the divisors of each number in the chain is the next number, and the sum of the divisors of the last number is the first; thus the chain is closed. For example,

the sum of the divisors of 12496 is 14288,
the sum of the divisors of 14288 is 15472,
the sum of the divisors of 15472 is 14536,
the sum of the divisors of 14536 is 14264,
the sum of the divisors of 14264 is 12496,

the first number of the chain. This is a well-known five-link chain. Two new four-link amicable chains,

(i) 2115324, 3317740, 3649556, 2797612,

and

(ii) 1264460, 1547860, 1727636, 1305184,

were found in 1965 by Dr. K. D. Fryer of the University of Waterloo. There is also a magnificent 28-link chain beginning with the number 14316. Also among Dr. Fryer's findings are examples of *semi-amicable pairs*, for which the divisor 1 is rejected; e.g.,

$$\text{(i)}\ (48, 75), \qquad \text{(ii)}\ (140, 195);$$

48 has divisors 2, 3, 4, 6, 8, 12, 16, 24 with sum 75, and 75 has divisors 3, 5, 15, 25 with sum 48.

Perfect Numbers. A positive integer n is said to be *perfect* if the sum of its proper divisors is n itself. Sometimes it is more convenient to use the equivalent definition that the sum of all divisors (including n itself) is $2n$. The smallest perfect number is $6 = 1 + 2 + 3$. The next four are 28, 496, 8128, and 33, 550, 336.

In 1975, only 24 perfect numbers are known, the last one, $2^{19936}(2^{19937} - 1)$, dating from 1971. Half were found since 1952 and, curiously, all 24 are even.

A truly remarkable formula for perfect numbers occurs in the writings of Euclid (and may have been known to the Pythagoreans). It was proved by Euclid that *every number of the form* $2^{n-1}(2^n - 1)$, *where* $2^n - 1$ *is a prime, is perfect.* Euler proved that, conversely, every *even* perfect number n is of this form. Before giving L. E. Dickson's elegant proof (which appeared in 1911) of Euler's theorem, let us re-state and prove Euclid's theorem:

If $2^n - 1$, $n > 1$, *is a prime, then the number*

$$m = 2^{n-1}(2^n - 1)$$

is perfect (and obviously even).

Proof: If $2^n - 1$ is a prime number p, then

$$m = 2^{n-1}(2^n - 1) = 2^{n-1}p.$$

The only possible divisors of m, then, are

$$1,\ 2,\ 2^2,\ 2^3,\ \cdots,\ 2^{n-1},\ p,\ 2p,\ 2^2p,\ 2^3p,\ \cdots,\ 2^{n-1}p.$$

We observe that the first n of these numbers form a geometric progression whose sum is $2^n - 1$, and that the remaining numbers form a geometric progression whose sum is $p(2^n - 1)$. Altogether, then, the sum of the divisors of m is

$$\begin{aligned}(2^n - 1) + p(2^n - 1) = (2^n - 1)(1 + p) &= (2^n - 1)(2^n) \\ &= 2[2^{n-1}(2^n - 1)] = 2m,\end{aligned}$$

as required. (Since we included the improper divisor m, itself, the sum should come out to be $2m$.)

Before proving the theorem of Euler, we need to derive a formula for the sum $\sigma(m)$† of all the divisors of m (including m itself). With the help of such a formula, we shall prove that the function $\sigma(m)$ is "multiplicative" in the sense that, if x and y are two positive integers having no common factor, then

$$\sigma(xy) = \sigma(x) \cdot \sigma(y).$$

To this end, let

$$m = p_1^{a_1} p_2^{a_2} \cdots p_k^{a_k}$$

represent the prime factorization of m. Then any divisor of m is of the form

$$p_1^{b_1} p_2^{b_2} \cdots p_k^{b_k} \quad \text{with} \quad 0 \le b_i \le a_i, \quad i = 1, 2, \cdots, k.$$

The products of this form as each exponent b_i takes on each permissible value $0, 1, 2, \cdots, a_i$ constitute all the divisors of m. Their sum $\sigma(m)$ can be obtained by expanding the product

$$(1 + p_1 + p_1^2 + \cdots + p_1^{a_1})(1 + p_2 + p_2^2 + \cdots + p_2^{a_2}) \cdots \\ \cdots (1 + p_k + p_k^2 + \cdots + p_k^{a_k}).$$

If we express each factor as the sum of a geometric series, we obtain the expression

$$\sigma(m) = \frac{p_1^{a_1+1} - 1}{p_1 - 1} \cdot \frac{p_2^{a_2+1} - 1}{p_2 - 1} \cdots \frac{p_k^{a_k+1} - 1}{p_k - 1}.$$

† The symbol σ is the lower case Greek letter "sigma"; we often encounter the capital sigma, Σ, symbolizing summation.

Now suppose that the integers x and y have no common factor, so that in their prime factorizations, say

$$x = p_1^{a_1}p_2^{a_2} \cdots p_k^{a_k}, \qquad y = q_1^{c_1}q_2^{c_2} \cdots q_n^{c_n},$$

none of the primes q_i are equal to any of the primes p_j. Then the prime factorization of xy is

$$xy = p_1^{a_1}p_2^{a_2} \cdots p_k^{a_k}q_1^{c_1}q_2^{c_2} \cdots q_n^{c_n},$$

and, by the above formula for $\sigma(m)$,

$$\begin{aligned} \sigma(xy) &= \frac{p_1^{a_1+1} - 1}{p_1 - 1} \cdots \frac{p_k^{a_k+1} - 1}{p_k - 1} \cdot \frac{q_1^{c_1+1} - 1}{q_1 - 1} \cdots \frac{q_n^{c_n+1} - 1}{q_1 - 1} \\ &= \sigma(x) \cdot \sigma(y), \end{aligned}$$

as we set out to show.

We now prove Euler's Theorem: *If m is an even perfect number, then*

$$m = 2^{n-1}(2^n - 1),$$

where $2^n - 1$ is a prime.

Proof: Since m is even, we may write it in the form

$$m = 2^{r-1}q,$$

where $r > 1$ and q is odd; i.e., 2^{r-1} is the largest power of 2 that divides m. If we could show that q has the value $2^r - 1$ and is a prime, then $m = 2^{r-1}(2^r - 1)$ would indeed be of the desired form.

In Dickson's elegant proof, both things are disposed of at once. Since 2^{r-1} and q have no common factors, we may use the multiplicative character of $\sigma(m)$ just derived to obtain

$$\sigma(m) = \sigma(2^{r-1}q) = \sigma(2^{r-1}) \cdot \sigma(q).$$

We evaluate the first factor on the right and obtain

$$\sigma(2^{r-1}) = 1 + 2 + 2^2 + \cdots + 2^{r-2} + 2^{r-1} = 2^r - 1,$$

so that

$$\sigma(m) = (2^r - 1)\sigma(q).$$

Since m is perfect, $\sigma(m) = 2m$. Thus

$$\sigma(m) = 2m = 2(2^{r-1}q) = 2^r q = (2^r - 1)\sigma(q).$$

Now every integer $t > 1$ has at least the divisors 1 and t. Therefore $\sigma(t) \geq 1 + t$ for all t, and the equality holds only if 1 and t are the only divisors of t, that is, if t is a prime.

In particular, we have

$$\sigma(q) \geq 1 + q$$

and should like to show that the equality holds. Let

$$\sigma(q) = d + q, \qquad d \geq 1;$$

we wish to show that $d = 1$, so that we can conclude that q is a prime.

From the above equation for $\sigma(m)$, $2^r q = (2^r - 1)\sigma(q)$, we obtain

$$2^r q = (2^r - 1)(q + d) = 2^r q - q + 2^r d - d,$$

yielding

$$q = 2^r d - d = d(2^r - 1).$$

Suppose d is not equal to 1. Then d must be greater than 1. But d cannot exceed $q/3$ because, in

$$q = d(2^r - 1),$$

r is at least 2, so that $q = d(2^r - 1) \geq d(2^2 - 1) = 3d$, and $d \leq q/3$. Thus d is different from both 1 and q. Yet, clearly, d is a divisor of q. Thus 1, d and q are among the divisors of q, so that $\sigma(q)$ is at least as big as $1 + d + q$; in symbols, $\sigma(q) \geq 1 + d + q$. So we have

$$\sigma(q) = d + q \geq 1 + d + q,$$

which is impossible. Hence $d = 1$, completing the proof.

Figurate Numbers. From Figure 13.1, it is clear why the numbers 1, 3, 6, 10, $\cdots$ are called triangular numbers. The formula for the n-th triangular number, t_n, is easily seen to be

$$t_n = \frac{n(n+1)}{2}$$

$$(= 1 + 2 + 3 + \cdots + n, \text{ counting the rows from the top}).$$

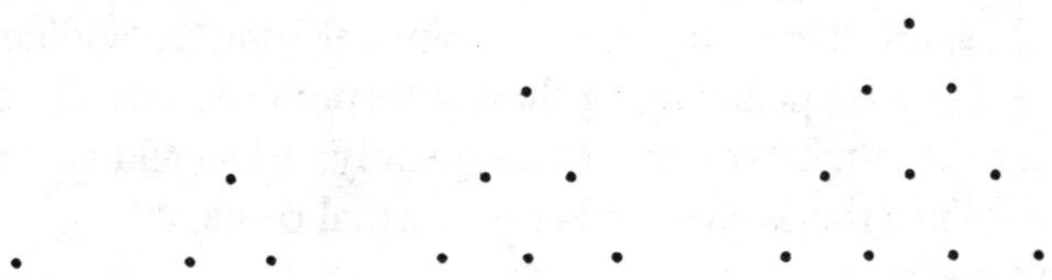

etc.

Figure 13.1

Similarly, the square numbers are

$$s_1 = 1,\ s_2 = 4,\ s_3 = 9,\ s_4 = 16,\ \cdots,\ s_n = n^2,\ \cdots.$$

The pentagonal numbers (see Figure 13.2) are $p_1 = 1$, $p_2 = 5$, $p_3 = 12$, $\cdots$,

$$p_n = \frac{n(3n-1)}{2}, \cdots.$$

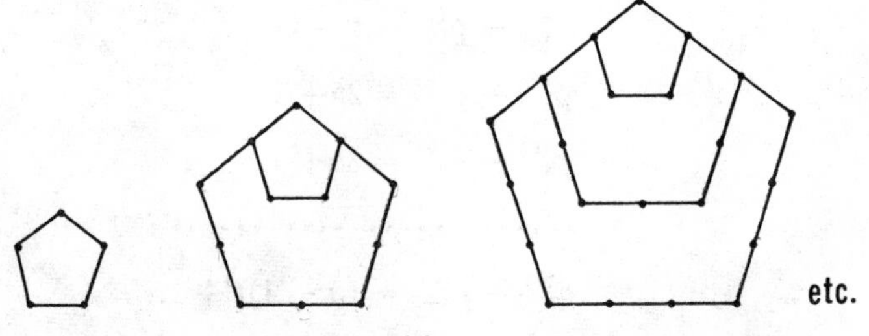

Figure 13.2

(We prove this formula later.) In general, there are the k-gonal numbers

$$p_1^{(k)} = 1,\ p_2^{(k)} = k,\ p_3^{(k)} = 3(k-1),\ \cdots,$$

$$p_n^{(k)} = \frac{n}{2}[n(k-2) - k + 4].$$

Let us prove this general formula for $p_n^{(k)}$, the n-th k-gonal number.

Looking carefully at Figure 13.2, we can see in general how the $(n+1)$-th k-gonal number is constructed from the n-th one. First of all, the figure for $p_{n+1}^{(k)}$ contains the figure for $p_n^{(k)}$.

We note that k is the number of sides and vertices in the overall figure,† and is constant; n is the number of points per side. The figure for $p_{n+1}^{(k)}$, then, has $n+1$ points along each side. To obtain the figure for $p_{n+1}^{(k)}$ from that for $p_n^{(k)}$ we must add $k-2$ sides. On each of these $k-2$ sides there are $n+1$ points. However, adding in $(k-2)(n+1)$ extra points along these sides provides a duplication at each of the "internal" corners. Consequently, we should subtract $k-3$ points, since this is the number of internal corners.

As a result, we have

$$p_{n+1}^{(k)} = p_n^{(k)} + (k-2)(n+1) - (k-3).$$

Setting $k-2=t$, we get

$$p_{n+1}^{(k)} = p_n^{(k)} + t(n+1) - (t-1),$$

or

$$p_{n+1}^{(k)} - p_n^{(k)} = tn + 1.$$

Thus

$$\begin{aligned}
p_2^{(k)} - p_1^{(k)} &= t + 1,\\
p_3^{(k)} - p_2^{(k)} &= 2t + 1,\\
p_4^{(k)} - p_3^{(k)} &= 3t + 1,\\
&\cdots\cdots\cdots\cdots\\
p_n^{(k)} - p_{n-1}^{(k)} &= (n-1)t + 1.
\end{aligned}$$

Adding up all these equations, we get

$$p_n^{(k)} - p_1^{(k)} = t[1 + 2 + 3 + \cdots + (n-1)] + (n-1),$$

so that

† Except in the case $n = 1$, where the k-gon reduces to a single point. The discussion is valid for all integers $n > 1$.

$$p_n^{(k)} = p_1^{(k)} + t \cdot \frac{(n-1)n}{2} + (n-1)$$

$$= 1 + t \cdot \frac{(n-1)n}{2} + n - 1 \quad (p_1^{(k)} = 1 \text{ for all } k)$$

$$= n\left[\frac{(n-1)t}{2} + 1\right] = \frac{n}{2}[nt - t + 2],$$

and after substituting $k - 2$ back in for t, we obtain

$$p_n^{(k)} = \frac{n}{2}[n(k-2) - k + 4],$$

which, for $k = 5$, yields the formula for pentagonal numbers on page 117.

From this formula one may deduce the surprising relation

$$p_n^{(k)} = p_n^{(k-1)} + p_{n-1}^{(3)}.$$

(See Exercise 1 at the end of this essay.)

We close this sub-section with a remarkable formula of Euler concerning the pentagonal numbers

$$p_n^{(5)} = \frac{n}{2}(3n-1): \quad 1,\ 5,\ 12,\ 22,\ 35,\ 51,\ 70,\ \cdots.$$

The numbers of the form

$$\frac{n}{2}(3n+1) \quad \text{are} \quad 2,\ 7,\ 15,\ 26,\ 40,\ 57,\ \cdots.$$

The elements of the set

$$\{1,\ 2,\ 5,\ 7,\ 12,\ 15,\ 22,\ 26,\ 36,\ 40,\ \cdots\},$$

obtained by alternating elements of these ordered sets, are called the *generalized pentagonal numbers*. Now Euler's formula says that the sum $\sigma(N)$ of *all* the divisors of N satisfies the relation

$$
\begin{aligned}
\sigma(N) - \sigma(N-1) &- \sigma(N-2) \\
+ \sigma(N-5) &+ \sigma(N-7) \\
- \sigma(N-12) &- \sigma(N-15) \\
+ \sigma(N-22) &+ \sigma(N-26) \\
- \quad \cdots \quad &- \quad \cdots \\
+ \quad \cdots \quad &+ \quad \cdots \\
\cdots\cdots\cdots&\cdots\cdots\cdots \\
= 0,&
\end{aligned}
$$

where the pairs of additions and subtractions of $\sigma(N - P_i)$, P_i the i-th generalized pentagonal number, are carried out so long as N is greater than or equal to P_i, and where $\sigma(0)$ is given the value N in the event that $\sigma(0)$ occurs.† For example, for $N = 15$,

$$
\begin{aligned}
\text{(i)}\quad \sigma(15) - \sigma(14) - \sigma(13)& \\
+ \sigma(10) + \sigma(8)& \\
- \sigma(3) \quad - \sigma(0)& = 24 - 24 - 14 \\
&\quad + 18 + 15 \\
&\quad - \ 4 - 15 \\
&= 0;
\end{aligned}
$$

$$
\begin{aligned}
\text{(ii)}\quad \sigma(27) - \sigma(26) - \sigma(25)& \\
+ \sigma(22) + \sigma(20)& \\
- \sigma(15) - \sigma(12)& \\
+ \sigma(5) \quad + \sigma(1)& = 40 - 42 - 31 \\
&\quad + 36 + 42 \\
&\quad - 24 - 28 \\
&\quad + \ 6 + \ 1 \\
&= 0.
\end{aligned}
$$

Pythagorean Triples. If the positive integers u, v, w represent the lengths of the sides of a right triangle, the trio (u, v, w) is called a *Pythagorean triple*. Let w denote the largest of the three numbers; then by the Theorem of Pythagoras

$$u^2 + v^2 = w^2. \tag{1}$$

† See the translation and discussion of Euler's Memoir in G. Pólya, *Mathematics and Plausible Reasoning*, Vol. I, *Induction and Analogy*, Princeton Univ. Press, 1954, Princeton, pp. 90–97.

For example, $(3, 4, 5)$ and $(5, 12, 13)$ are Pythagorean triples.

We observe that any multiple of a triple yields another triple: if $u^2 + v^2 = w^2$, then

$$(ku)^2 + (kv)^2 = k^2(u^2 + v^2) = (kw)^2;$$

thus $(3, 4, 5)$, for example, leads to $(6, 8, 10)$, $(9, 12, 15)$, and so on.

Triples such that the three numbers have no common factor are called *primitive* triples. *In a primitive triple, no two of the three numbers can have a common factor*, because if they did, the relation (1) would force the third to have the same factor. For example, if

$$u = du_1, \quad w = dw_1, \qquad \text{then} \qquad v^2 = w^2 - u^2 = d^2(w_1^2 - u_1^2),$$

so that d is a factor of v also. That is to say, there are only two kinds of triples:

(i) those in which all three numbers have a common factor;

(ii) those in which no two of the numbers have a common factor.

Consequently, there exist only primitive triples and multiples of primitive triples. Thus the problem of finding all Pythagorean triples is reduced to that of finding the primitive triples, from which all others arise.

Before stating and proving the theorem that allows us to generate all primitive Pythagorean triples, we list some additional properties of such triples.

Since no two of the three numbers have a common factor it follows, in particular, that *u and v are not both even.*

We observe, moreover, that *u and v cannot both be odd*; for, if $u = 2s + 1$, $v = 2t + 1$, then

$$w^2 = (2s + 1)^2 + (2t + 1)^2 = 4(s^2 + t^2 + s + t) + 2$$

so that w^2 would be divisible by 2, but not by 4. This cannot be (because, if w^2 is divisible by 2, w is even and so w^2 must be divisible by 4).

In what follows we shall denote a primitive triple by (x, y, z), where x is even, y is odd, so z is odd; say

$$x = 2a, \qquad y = 2b + 1, \qquad \text{and} \quad z^2 = x^2 + y^2.$$

THEOREM. *Every primitive Pythagorean triple* $x,\ y,\ z$ *is of the form*

$$(x,\ y,\ z) = (2mn,\ m^2 - n^2,\ m^2 + n^2) \tag{2}$$

where

(a) m *and* n *have no common factor, and*
(b) *one of* $m,\ n$ *is even, the other odd.*

Conversely, if m *and* n *satisfy the conditions* (a) *and* (b), *then* (2) *is a primitive triple.*

The identity

$$(2mn)^2 + (m^2 - n^2)^2 = (m^2 + n^2)^2$$

guarantees that the triple (2) is indeed Pythagorean. So, to prove the theorem, we need to show that

I. if $(x,\ y,\ z)$ is primitive with x even, then it is of the form (2), where m and n have properties (a) and (b);

II. if $m,\ n$ have properties (a) and (b), then the triple $(x,\ y,\ z)$ given by (2) is primitive.

Proof: I. We have seen that, if (x, y, z) is primitive and x even, then y and z are odd. Therefore

$$z + y = 2k, \qquad z - y = 2l \tag{3}$$

are even, and, from (3),

$$z = k + l, \qquad y = k - l. \tag{4}$$

Since $x = 2a$ satisfies

$$x^2 = 4a^2 = z^2 - y^2 = (z + y)(z - y) = 4kl,$$

we conclude that $a^2 = kl$. Now k and l can have no common factor; for, if they did, y and z would have the same common factor, by (4), and this would contradict our assumption that $(x,\ y,\ z)$ is primitive. Thus kl can be a perfect square only if k and l both are perfect squares:

$$k = m^2, \qquad l = n^2,$$

with m and n having no common factor.† Thus m, n have property (a), and $x^2 = 4m^2n^2$ implies that

$$x = 2mn, \qquad y = k - l = m^2 - n^2, \qquad z = k + l = m^2 + n^2.$$

Since m and n have no common factor, they are not both even; moreover, they are not both odd because, if they were, $y = m^2 - n^2$ and $z = m^2 + n^2$ would both be even, a contradiction. Thus m, n also satisfy condition (b).

II. If m, n have no common factor, neither do $k = m^2$ and $l = n^2$; if precisely one of the numbers m, n is even, then precisely one of the numbers k, l is even and

$$y = k - l, \qquad z = k + l$$

are both odd. If y and z had a common factor, it would have to be odd, and the numbers

$$z + y = 2k, \qquad z - y = 2l$$

would have that odd factor in common, contradicting the first assertion of this paragraph. Thus y, z have no common factor.

We now set $x = 2mn$ and obtain

$$x^2 + y^2 = 4m^2n^2 + (m^2 - n^2)^2 = (m^2 + n^2)^2 = z^2,$$

from which we conclude that x, y, z is primitive; for, if x and y or x and z had a common factor, so would y and z, contrary to the result just derived. This concludes the proof that all primitive Pythagorean triples are generated by (2), with m, n satisfying conditions (a) and (b), and that only such triples are so generated.

We close this section by stating a remarkable property of Pythagorean triples:

† If m and n had a common factor, k and l would also have that factor.

In a primitive triple (x, y, z), *either* x *or* y *is divisible by* 3; *either* x *or* y *is divisible by* 4; *and either* x, y, *or* z *is divisible by* 5.

For example, in the triple (8, 15, 17) 15 is divisible by 3, 8 by 4, 15 by 5.

For a method of generating all primitive Pythagorean triples with the additional property that x, y are consecutive integers, see Problem 3 of Problem Set 20 in C. D. Olds, *Continued Fractions*, (NML Vol. 9) Random House, 1963, New York, p. 121, and its solution on pp. 154–155.

EXERCISES

1. Verify the relation $p_n^{(k)} = p_n^{(k-1)} + p_{n-1}^{(3)}$.

2. From the representation (2) and the theorem, conclude that x is divisible by 4.

3. Prove that either x or y is divisible by 3. [*Hint*: If either m or n is divisible by 3, then $x = 2mn$ is divisible by 3; so it suffices to show that, if neither m nor n is divisible by 3, then $y = m^2 - n^2$ is divisible by 3. To see this, observe that, for any integer N not divisible by 3, we have either $N \equiv 1 \pmod 3$ or $N \equiv -1 \pmod 3$, and in both cases $N^2 \equiv 1 \pmod 3$.]

4. Prove that one of the numbers x, y, z is divisible by 5. [*Hint*: Use a modification of the hint to the previous exercise. In particular, observe that a number M not divisible by 5 satisfies one of the four congruences $M \equiv \pm 1 \pmod 5$, $M \equiv \pm 2 \pmod 5$].

5. Prove: If $2^q - 1$ is a prime, then q is a prime.

6. Prove: An even perfect number ends in a 6 or an 8.

REFERENCES

H. Eves, *Introduction to the History of Mathematics*, (3rd ed.) Holt, Rinehart and Winston, 1969, New York.

H. N. Wright, *First Course in Theory of Numbers*, Wiley, 1939, New York.

ESSAY FOURTEEN

Abundant Numbers

In our work on Pythagorean Arithmetic we encountered "perfect numbers", that is, integers n whose divisors add up to $2n$ [e.g., 6, for $\sigma(6) = 1 + 2 + 3 + 6 = 12$]. If the sum of the divisors, $\sigma(n)$, of a positive integer n is *less than* $2n$, the number is said to be *deficient*; and if $\sigma(n)$ *exceeds* $2n$, the number is called *abundant*. For example,

$$10 \text{ is deficient—}\sigma(10) \text{ is only } 18,$$

while

$$12 \text{ is abundant—}\sigma(12) \text{ is } 28.$$

Clearly every positive integer is either deficient, perfect, or abundant.

These definitions seem to provide so little for us to work on that we are quite surprised to find easy proofs for such general results as the following two theorems, our main concern in this essay.

THEOREM 1. *Every even number greater than* 46 *can be expressed as the sum of two abundant numbers.*

THEOREM 2. *Every integer greater than or equal to* 83,160 *can be expressed as the sum of two abundant numbers.*

Proof of Theorem 1: In the proofs of both these theorems, a fundamental role is played by the lemma:

LEMMA. *If n is either perfect or abundant, then its multiples mn $(m = 2, 3, 4, \cdots)$ are abundant.*

Proof of Lemma: Let the divisors of n be $d_1 = 1, d_2, \cdots, d_k = n$. Then $md_1, md_2, \cdots, md_k$ are divisors of the number mn. Since $m \geq 2$, the divisor 1 of mn is not included in this list. Without even bothering to look for other divisors of mn, we may conclude that

$$\begin{aligned}\sigma(mn) &\geq 1 + md_1 + \cdots + md_k \\ &> md_1 + md_2 + \cdots + md_k = m\sigma(n).\end{aligned}$$

Now if n is perfect or abundant, then $\sigma(n) \geq 2n$. Hence

$$\sigma(mn) > m(2n)$$

implying that mn is abundant.

Now it is a simple matter to complete the proof of Theorem 1. Clearly, when an even integer is divided by 6, it has either the remainder 0, or 2, or 4; i.e., every even integer is of one of the forms

$$n = 6k, \quad \text{or} \quad n = 6k + 2, \quad \text{or} \quad n = 6k + 4.$$

If $n = 6k$ and $k \geq 4$, we may write

$$n = 6k = 6 \cdot 2 + 6(k - 2)$$

and, by our lemma, each term on the right is abundant.

If $n = 6k + 2$ and $k \geq 5$, we may write

$$n = 6k + 2 = 6 \cdot 3 + 6(k - 3) + 2 = 20 + 6(k - 3);$$

20 is abundant and so is the multiple of 6 in the second term.

If $n = 6k + 4$ and $k \geq 8$, we may write

$$n = 6k + 4 = 6 \cdot 6 + 6(k - 6) + 4 = 40 + 6(k - 6);$$

40 is abundant and so is $6(k - 6)$.

The abundant numbers less than 46 are 12, 18, 20, 24, 30, 36, 40 and 42. It is easily checked that no two of these add up to 46. Consequently, 46 is the largest even integer which cannot be written as a sum of two abundant numbers.

Proof of Theorem 2: By Theorem 1, every even integer greater than or equal to 83,160 can be expressed as the sum of two abundant numbers. We need consider, then, only the odd numbers exceeding 83,160.

Our basic approach is the same as it was in Theorem 1: we shall actually find a form $A + B$ which displays every odd number exceeding 83,160 (indeed, every even number too) as the sum of two abundant numbers. Again, the abundance of the summands will result from the fact that they are multiples of known abundant numbers.

Clearly if $A + B$ is to give an odd sum, A and B must be of opposite parity, that is, one must be odd and the other even. Consequently, we begin by looking for an odd abundant number. It turns out that we must look for a long time; the smallest odd abundant number is 945. (The next is 1575.) This is not very encouraging, for multiples of 945 don't give a very dense covering of the integers.

However, there is a result from the field of Diophantine Equations which comes to our rescue:

If a and b are relatively prime positive integers, the equation $ax + by = c$, where $c > ab$ has a solution (x, y) in positive integers.

Accepting this for the moment, we look for an even abundant number which is relatively prime to $945 = 3^3 \cdot 5 \cdot 7$. The smallest such number is $88 = 2^3 \cdot 11$. The equation

$$88x + 945y = c$$

has a solution in *positive* integers (x, y) for every value of c exceeding $88 \cdot 945 = 83{,}160$. Since 88 and 945 are both abundant, $88x$ and $945y$ are abundant even when x or y is as small as 1; that is, any positive x, y make them abundant. And so our theorem follows.

Since 88 is the smallest abundant or perfect number coprime to 945, the number 83,160 is the smallest bound we can obtain by this method.† It has been shown that the largest odd number which cannot be written as the sum of two abundant numbers is 20161.

We conclude by establishing our Diophantine result.

Consider the equation

$$ax + by = d,$$

where d is an arbitrary positive integer. If even one integral solution exists, then there are infinitely many. For, suppose (x_0, y_0) is a known solution; i.e., $ax_0 + by_0 = d$. Then for any integer t, the integers

$$x = x_0 - bt, \qquad y = y_0 + at$$

also constitute a solution because

$$ax + by = a(x_0 - bt) + b(y_0 + at) = ax_0 + by_0 = d.$$

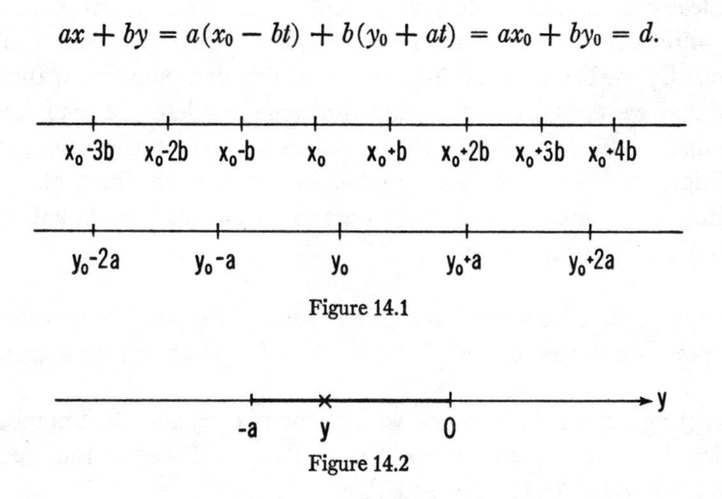

Figure 14.1

Figure 14.2

Now the values of x in these solutions occur on a number axis at intervals of b units; the values of y at intervals of a units (see Figure 14.1). Consequently, if there is a solution at all, there is one whose value of y lies in the half-closed interval $-a < y \leq 0$ (often denoted

† The Diophantine result can be sharpened somewhat to yield: every integer greater than or equal to $(88 - 1)(945 - 1) = 87 \cdot 944 = 82{,}128$ can be written as a sum of two abundant numbers.

by $(-a, 0]$; see Figure 14.2). Let us denote such a solution by (x', y'); that is,

$$ax' + by' = d, \qquad \text{where } y' \text{ is in } (-a, 0].$$

Since $y' \leq 0$, and a, b, and d are positive, it follows that $x' > 0$. Also, since y' lies in $(-a, 0]$, the number $y' + a$ is positive. As a result,

$$(ax' + by') + ab = d + ab$$

gives

$$ax' + b(y' + a) = d + ab,$$

where both x' and $(y' + a)$ are *positive* numbers. That is to say, *if* the equation $ax + by = d$ $(d > 0)$ has any integral solution at all, then the corresponding equation $ax + by = ab + d$ has a solution in *positive* integers (x, y).

But, for a and b relatively prime, we have, by the famous Euclidean Algorithm, that the equation $ax + by = 1$ has an integral solution.†

Consequently the equation

$$ax + by = d \qquad (d \text{ any positive integer})$$

has a solution (dx, dy), where (x, y) is a solution of $ax + by = 1$. Thus the equation

$$ax + by = c \qquad (\text{where } c > ab)$$

has a solution in positive integers. ($c = ab + d$ for some positive d.)

† For an exposition of Euclid's Algorithm see, for example, C. D. Olds, *Continued Fractions*, (NML Vol. 9), Random House, Inc., 1963, New York p. 17.

ESSAY FIFTEEN

Mascheroni and Steiner

The Euclidean tools for carrying out geometric constructions were straightedge and compasses. We need to distinguish between straightedge and ruler. A *ruler* has markings on it which permit its use for transporting lengths from one place to another; moreover, it has two edges permitting certain parallel lines to be drawn. A *straightedge*, on the other hand, simply enables us to join two given points by a straight line. Euclidean compasses also are to be distinguished from dividers; compasses may be used only to draw circles with a given centre A which pass through another given point B (i.e., of radius AB). The modern practice of such maneuvers as "with centre A, radius CD (a transported radius), construct a circle $\cdots$" are not in keeping with Euclid's use of compasses; he presumed that they would collapse when either arm was lifted from the page. It turns out, however, that the difference between collapsing and "divider-type" compasses is only apparent, for we shall soon show that Euclid's compasses can achieve any construction executed by its modern counterpart.

We are chiefly concerned in this essay with two major discoveries regarding the equivalence of instruments. In 1797, the Italian geometer Lorenzo Mascheroni showed that any construction which can be carried out with straightedge and compasses can be carried out with compasses alone! Following the suggestion of J. V. Poncelet, Jakob Steiner proved in 1833 that any construction that can be executed with straightedge and compasses can be carried out with straightedge alone, provided that just one circle and its centre are given.

Of course, two obvious conventions must be made in all this work:

(I) in constructions involving just compasses, we consider a required straight line as obtained when two points on it are determined;

(II) in constructions involving just straightedge, a required circle is considered obtained when its centre and radius are determined.

We interject here, before giving our proofs, an interesting historical note. Shortly before 1928, a student of the Danish mathematician J. Hjelmslev, while browsing in a bookstore in Copenhagen, came across a copy of an old book, *Euclides Danicus*, published in 1672 by an obscure writer named Georg Mohr. Upon examining the book, Hjelmslev was surprised to find that it contained Mascheroni's discovery, with a proof, arrived at 125 years before Mascheroni's publication. Now on to our propositions.

(i) The Collapsing and Modern Compasses are Equivalent

All we need to show here is that a circle with centre A and (transported) radius BC can be constructed with collapsing compasses. We denote by $A(AB)$ the circle with centre A and radius AB. Each circle in the following construction can be drawn with collapsing compasses.

Construction. In Figure 15.1, A is a given point and BC a given segment. First we draw $A(AB)$ and $B(BA)$ and denote their intersections by D and E. Then we draw $D(DC)$ and $E(EC)$ to meet again at X. We claim that then $AX = BC$, making $A(AX) = A(BC)$. (Actually we have reflected the point C in the line through D and E.)

Proof of claim: Since $DX = DC$ and $EX = EC$, triangles DXE and DCE are congruent. Since $AD = AE(=AB) = BD = BE$, triangles DAE and DBE are congruent. Consequently

$$\measuredangle XDE = \measuredangle CDE, \qquad \measuredangle ADE = \measuredangle BDE,$$

and so their differences are equal, that is,

$$\measuredangle XDA = \measuredangle CDB.$$

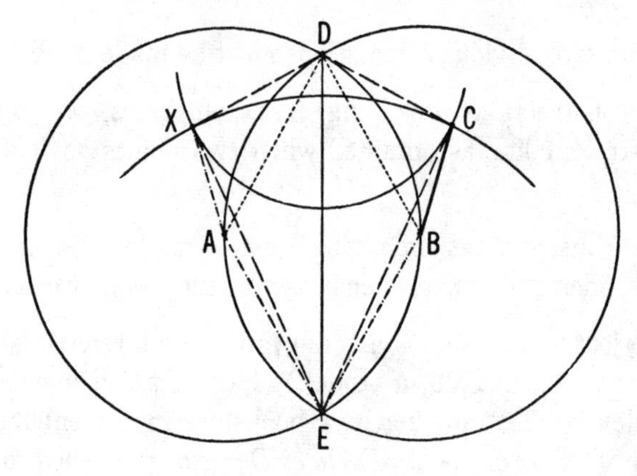

Figure 15.1

But $DX = DC$, and $DA = DB$, so that triangles XDA and CDB are congruent. Hence $AX = BC$ (QED).

All constructions with straightedge and compasses ultimately depend upon finding

(a) the points of intersection of two circles;
(b) the point of intersection of two straight lines;
(c) the points of intersection of a straight line and a circle.

Obviously the first is easily accomplished with compasses alone, the second with straightedge alone. Mascheroni's proposition, then, is established upon showing how to construct (b) and (c) with compasses only; Steiner's upon showing how to construct (a) and (c) with straightedge only.

(ii) Mascheroni: Constructions With Compasses Alone

We begin by establishing four preliminary constructions.

1. *To construct the mirror image Y of a point X with respect to a line PQ.*

Draw the circles $P(PX)$ and $Q(QX)$; their second point of intersection is the required point Y (see Figure 15.2).

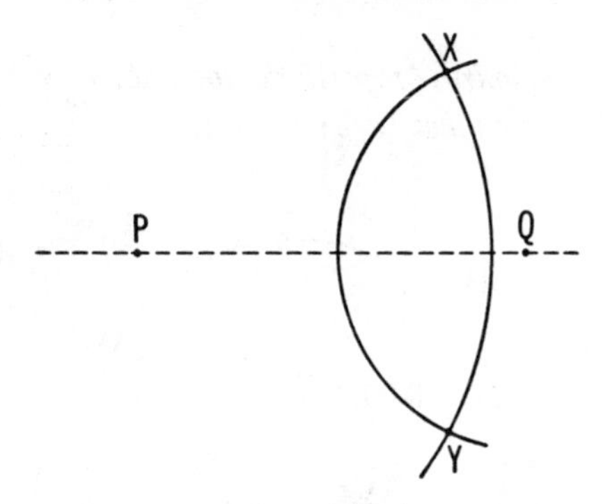

Figure 15.2

Proof: To show that X and Y are mirror images of each other in line PQ, we must show that PQ is the perpendicular bisector of XY. By construction, $PX = PY$ and $QX = QY$; thus P and Q lie on the perpendicular bisector of XY.

2. *To construct a segment* n *times as long as a given segment* AB; *i.e., to construct* nAB, n *a positive integer.*

Step off the radius AB around the circumference of $B(BA)$, beginning at A to produce C, D, E (see Figure 15.3). Then ABE is a straight line and $AE = 2AB$. Repeat this procedure to build up segments of lengths $3AB$, $4AB$, $\cdots$, nAB.

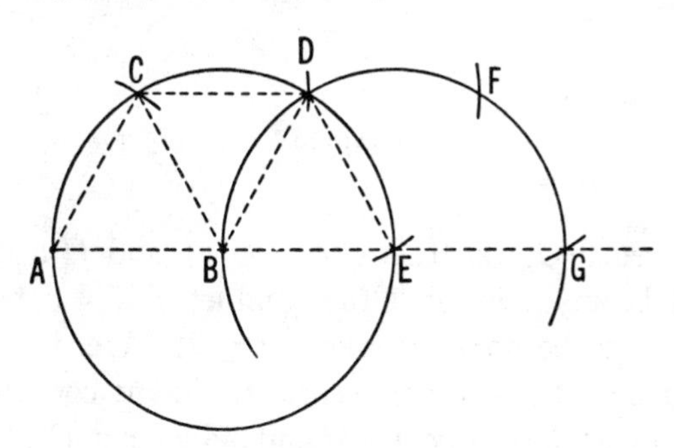

Figure 15.3

Proof: $AB = AC = CB = CD = BD = BE = DE$, all equal to radius AB. This means that triangles ABC, BCD, BDE are equilateral, so that each of the angles ABC, CBD, DBE is 60°. Therefore ABE is a straight line, and its length is $AE = 2AB$.

3. *To construct the fourth proportional to* a, b, c; *i.e., given* a, b, c, *to construct* x *such that* $(a/b) = (c/x)$.

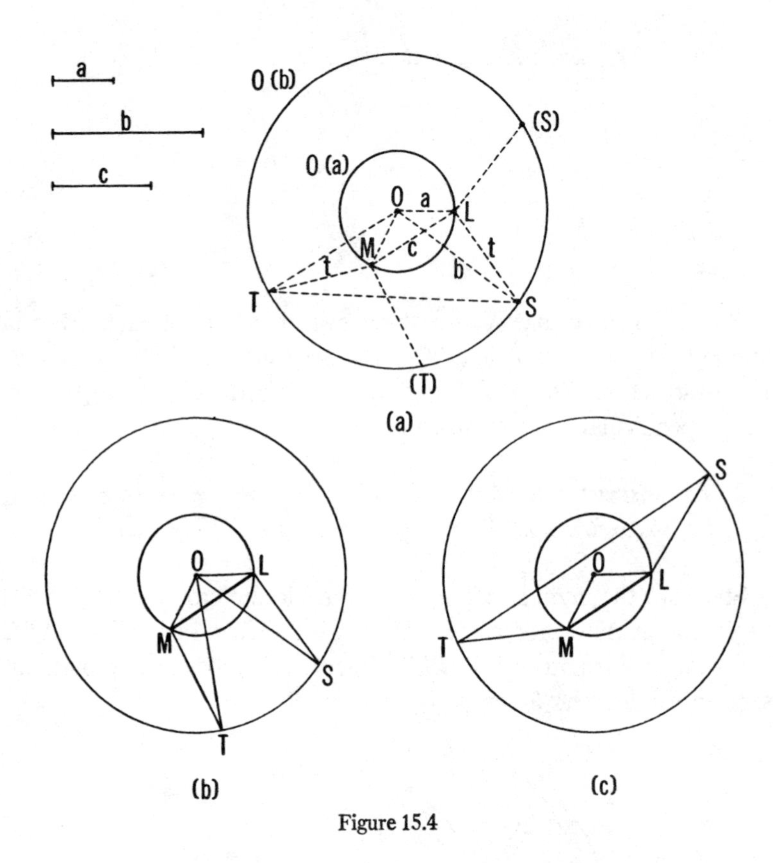

Figure 15.4

With any centre O, construct circles $O(a)$ and $O(b)$ (see Figure 15.4). Let L be any point on $O(a)$, and let $LM = c$ be a chord of $O(a)$. (This can be done whenever $c \leq 2a$. We discuss the case $c > 2a$ in a moment.) Now with centre L and any convenient radius, say t, cut $O(b)$ at S. With centre M and radius t cut $O(b)$ again at T.

Now there are two possible positions for S on $O(b)$ and two for T. We are to select positions for S and T so that exactly one of OS, OT lies in the angle LOM [as in Figure 15.4(a) and *not* as in Figures 15.4(b) and (c)].

Clearly $\triangle LOS \cong \triangle MOT$ (corresponding sides are equal), giving $\sphericalangle LOS = \sphericalangle MOT$. Consequently, $\sphericalangle LOM = \sphericalangle SOT$. Thus isosceles triangles LOM and SOT are similar, and therefore

$$\frac{OL}{OS} = \frac{LM}{ST}, \qquad \text{that is,} \qquad \frac{a}{b} = \frac{c}{ST},$$

making ST the required fourth proportional.

This construction must be modified if c is greater than the diameter of $O(a)$, i.e., if $c > 2a$. Now we want x such that $(a/b) = (c/x)$, but this is the same x for which $(na/nb) = (c/x)$. By taking n so big that $2na > c$ (such an n always exists by Archimedes' axiom), we find na and nb by construction 2 above, and then proceed with the construction just explained, using na and nb instead of a and b. In this case the point M will certainly occur on $O(na)$, and the above procedure gives the required fourth proportional.

4. *To find the midpoint F of an arc AB whose center O is known.*

Let $A(AO)$ meet $O(AB)$ at C (Figure 15.5), and let $B(OB)$ meet $O(AB)$ at D. Then $ACOB$ and $BDOA$ are congruent parallelograms (pairs of opposite sides are equal) in which diagonals CB and DA are equal. Let $C(CB)$ meet $D(DA)$ at E. Let $C(OE)$ meet the given arc at F. Then F is the required point.

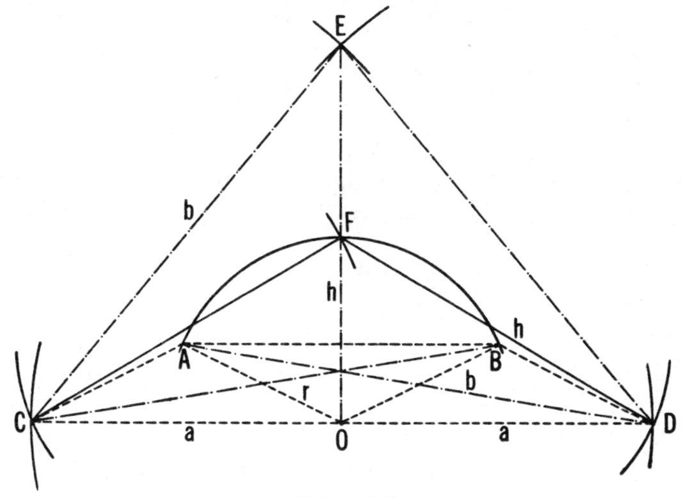

Figure 15.5

Proof: We can establish that F is the midpoint of the arc AB by showing that the radius OF is perpendicular to the chord AB. (The perpendicular from the centre of a circle to a chord bisects the chord and the arc cut off by the chord.) For convenience, set $r = OA$, $a = OC$, $b = AD$ and $h = OE$.

Now COD is a straight line because OC and OD are both parallel to AB. We know that EO is perpendicular to COD (indeed, EO is the perpendicular bisector of COD since O is its midpoint and $EC = ED$). But we don't know that F lies on EO. However, in right triangle ECO, $CE^2 = EO^2 + CO^2$, that is, $b^2 = h^2 + a^2$. Now we apply to $BDOA$ the theorem of elementary geometry that says "the sum of the squares of the diagonals of a parallelogram is equal to the sum of the squares of its sides":

$$AD^2 + OB^2 = 2AO^2 + 2DO^2,$$

i.e.,

$$b^2 + r^2 = 2r^2 + 2a^2, \qquad (OB = AO = r),$$

giving

$$b^2 = r^2 + 2a^2.$$

Thus,

$$h^2 + a^2 = r^2 + 2a^2,$$

whence

$$h^2 = r^2 + a^2.$$

This means that $\measuredangle COF$ is a right angle by the converse of the theorem of Pythagoras applied to $\triangle CFO$ (QED).

We may proceed now with the main problems (b) and (c) as follows [we take (c) first]:

(c) *To determine the intersection of a circle* $O(r)$ *and a straight line* AB.

By 1, determine O', the mirror image of O in AB. Then, if the intersection exists, i.e., if r is not less than the distance from O to line AB, $O'(r)$ meets $O(r)$ in the required points M and N [see Figure 15.6(a)].

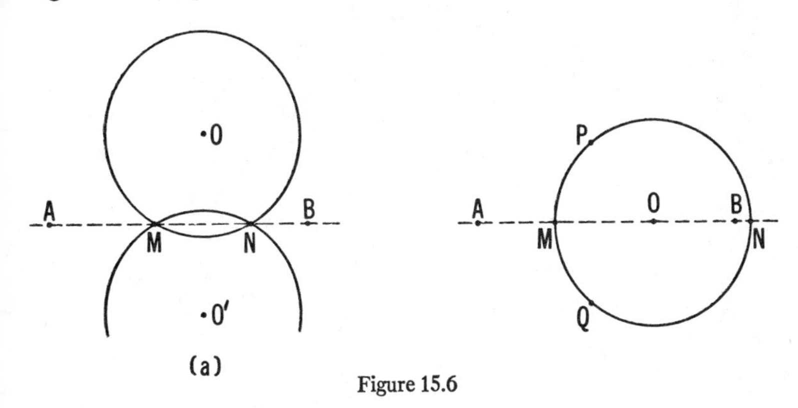

Figure 15.6

Proof: Since O' is the mirror image of O in AB, AB is the perpendicular bisector of OO'. But, clearly $MO = MO' = NO = NO' = r$, so M and N lie on the perpendicular bisector of OO', i.e., on AB.

There is a case, however, when this construction fails: if O lies on AB [see Figure 15.6(b)]. In this case, O' coincides with O (informing us of the special circumstances). We proceed simply to choose any point P on the circle $O(r)$ and find its mirror image Q in AB. Q also will lie on the given circle, and the required points M and N will be the midpoints of the arcs PQ (which we determine by 4 above). (If Q coincides with P, then $P = M$ and we step off the radius around to N as in 2 above.)

(b) *To find the intersection of two straight lines* AB, CD.

This time we prove things as we go along (see Figure 15.7). By 1, construct the mirror images C', D' of C, D in AB. Complete parallelogram $CC'ED$ by drawing $C'(CD)$ and $D(CC')$. (DD' is parallel to CC', both being perpendicular to AB. But DE is parallel to CC'; hence $ED'D$ is a straight line.)

Now the line $C'D'$, the image of line CD in AB, meets CD at F on AB (all lines meet their images on the mirror). Then because FD is parallel to $C'E$, triangles $D'FD$ and $D'C'E$ are similar, so that

$$\frac{D'E}{D'D} = \frac{D'C'}{D'F}.$$

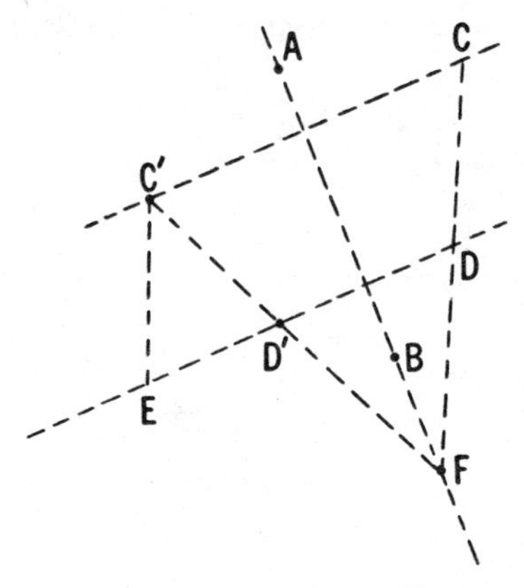

Figure 15.7

Now $D'E$, $D'D$, $D'C'$ are all known. Hence, by 3 above, we can construct the fourth proportional to these quantities to give the length of $D'F$; call it k. But $FD' = FD$ (F is on the perpendicular bisector of DD'). Hence circles $D(k)$ and $D'(k)$ meet at F, the required point.

This completes the proof of Mascheroni's proposition. We proceed now with Steiner's proposition.

(iii) Steiner: Constructions with Straightedge Only (Given a Fixed Circle with Its Centre)

Again some preliminary constructions are needed.

1. *To construct through a point P a line parallel to a given line AB.*

This problem breaks down into two cases.

(a) If the midpoint C of AB is known, we proceed as follows (see Figure 15.8): Take any point T on AP produced and join it to B and C. Let PB meet TC at S, and let AS meet TB at Q. Then PQ is the required parallel.

Proof: We are given that $AC = CB$ and have to prove that $PQ \parallel AB$.

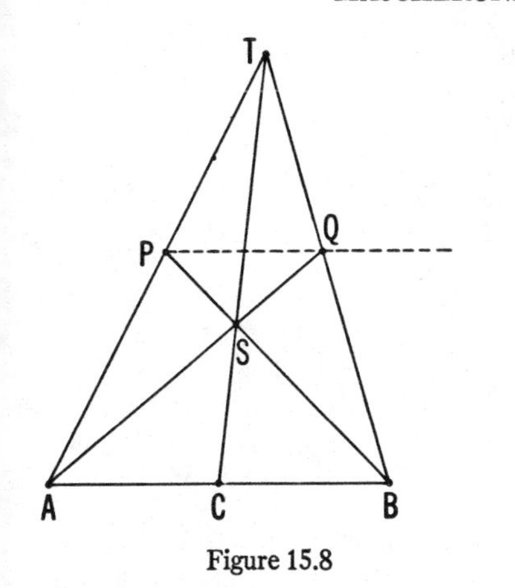

Figure 15.8

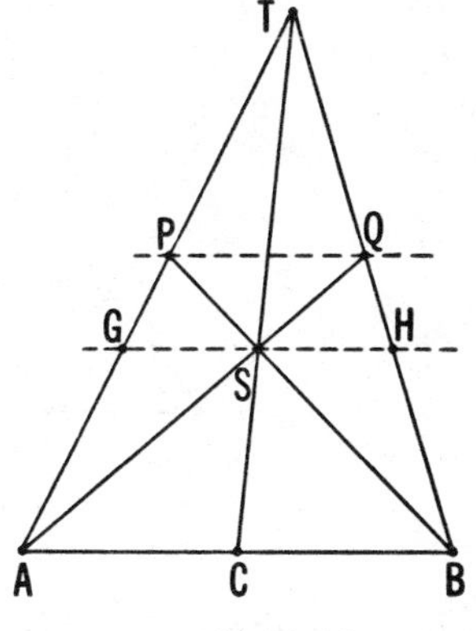

Figure 15.9

Imagine line GH drawn through S parallel to AB (see Figure 15.9). Then $\triangle TGS \sim \triangle TAC$, and $\triangle TSH \sim \triangle TCB$ so that

$$\frac{GS}{AC} = \frac{TS}{TC} = \frac{SH}{CB}.$$

Since $AC = CB$, it follows that $GS = SH$. Now since

$$\triangle PGS \sim \triangle PAB \qquad \text{and} \qquad \triangle QSH \sim \triangle QAB,$$

we have

$$\frac{PB}{PS} = \frac{AB}{GS} = \frac{AB}{SH} = \frac{QA}{QS};$$

hence

$$\frac{PB}{PS} - 1 = \frac{QA}{QS} - 1, \qquad \text{that is,} \qquad \frac{PB - PS}{PS} = \frac{QA - QS}{QS},$$

from which we have

$$\frac{SB}{PS} = \frac{SA}{QS}$$

(see Figure 15.9). Thus $\triangle SPQ \sim \triangle SBA$ (two sets of sides being proportional and the included angles equal), so it follows that $\measuredangle SPQ = \measuredangle SBA$. This implies that PQ is parallel to AB.

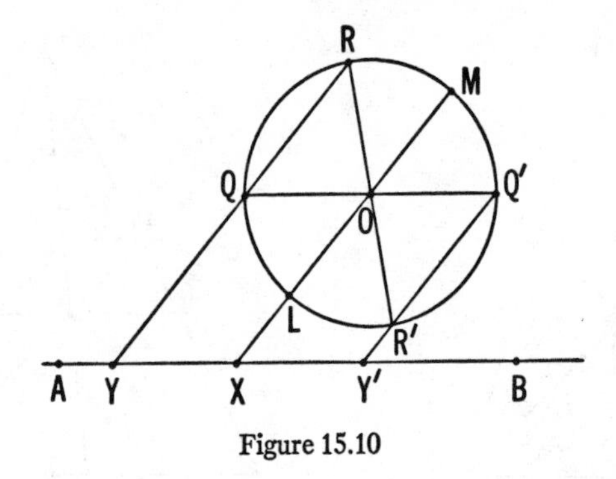

Figure 15.10

(b) If the midpoint of AB is not known, we proceed as follows (see Figure 15.10). Join any point X on AB to O, the centre of the fixed given circle, to yield diameter LOM. Then O is the midpoint of LM and, choosing R anywhere on the circle (except at L or M), we can, by the above method, construct RQ parallel to $MOLX$ to meet AB in Y.

The diameters through R and Q are drawn to give points R' and Q' on the circle. Let $Q'R'$ meet AB in Y'. Then it turns out (the easy proof is left to the reader) that X is the midpoint of YY' on AB. Hence the required parallel may now be constructed as above.

Now we come to the easy problem:

2. *To construct a line through P perpendicular to a line* AB.

Through a point Q on the fixed circle (see Figure 15.11) draw QR parallel to AB (choose Q so that QR is *not* a diameter). Then draw diameter QO to meet the circle again in Q'. Then PC, parallel to $Q'R$, is the required perpendicular.

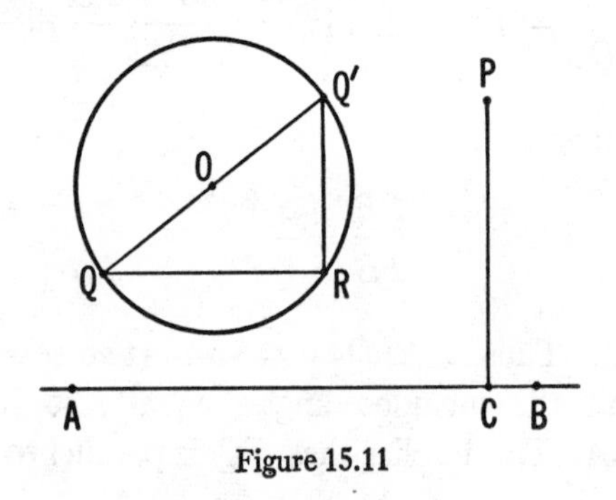

Figure 15.11

Proof: QR is parallel to AB and $Q'R$ is parallel to PC. Thus $\sphericalangle QRQ' = \sphericalangle ACP$. But $\sphericalangle QRQ'$ is a right angle (inscribed in a semicircle). Hence $\sphericalangle ACP$ is a right angle, as required.

Next we need the fundamental construction:

3. *To mark off a length* $XY = k$ *along a line* AB *in both directions from a point* P.

Segment XY is given; also line AB with P on it and a fixed circle with centre O are given (see Figure 15.12).

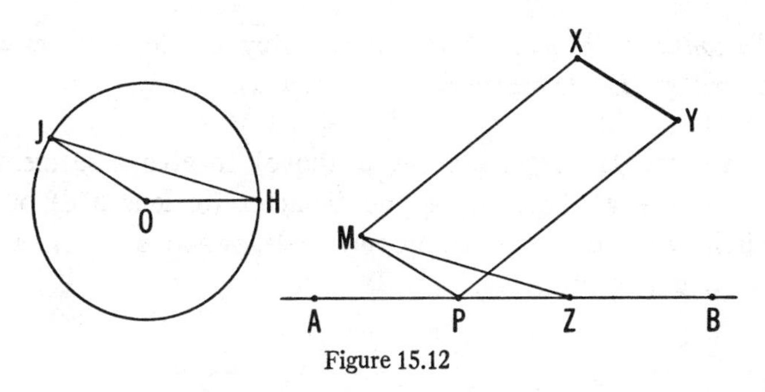

Figure 15.12

Let us mark off $PZ = k$ toward B. (The case "toward A" is similar.) First we draw PM and XM parallel to XY and PY, respectively, to complete parallelogram $XMPY$. (This gives $XY = PM$.) Now draw radii OH and OJ of the fixed circle parallel to AB, PM, respectively. Finally, draw MZ parallel to JH; then $PZ = k$.

Proof: The three pairs of parallels (OJ, PM), (OH, PZ), (JH, MZ) make the angles of $\triangle PMZ$ equal to those of $\triangle OJH$. But $\triangle OJH$ is isosceles, implying that $\triangle PMZ$ is, too. Thus $PZ = PM = XY$.

We consider next the easy problem:

4. *To construct the fourth proportional to* a, b, c.

On any two intersecting straight lines s and t we lay off segments a, b, c as shown in Figure 15.13. (Construction 3 is needed to do this.) Drawing MN parallel to PQ gives QN, the required fourth proportional. (The easy proof is omitted.)

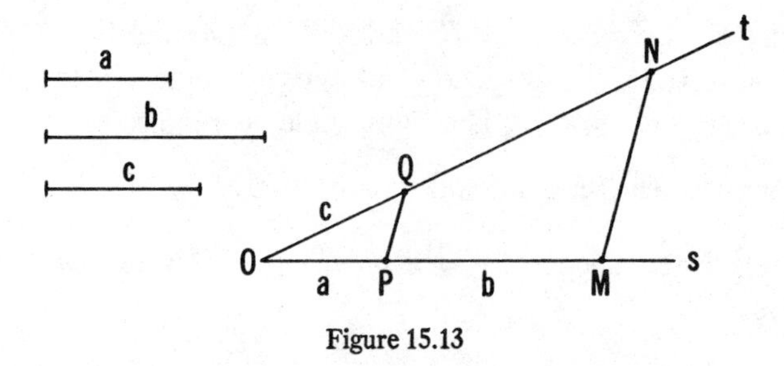

Figure 15.13

Finally we show how

5. *To construct the mean proportional* $\sqrt{xy}$ *of two given segments* x *and* y; *i.e., to construct* z *such that* $x/z = z/y$.

First we add the segments (by 3 above) to give a segment of length $t = x + y$. Then, using the diameter (of length d) of the fixed circle, we find the fourth proportionals, m and n, of (t, x, d) and (t, y, d), respectively.
Since

$$\frac{t}{x} = \frac{d}{m} \quad \text{and} \quad \frac{t}{y} = \frac{d}{n}, \qquad \text{or} \qquad m = \frac{dx}{t}, \quad n = \frac{dy}{t},$$

we have

$$m + n = \frac{dx + dy}{t} = \frac{d(x + y)}{t} = \frac{dt}{t} = d.$$

Now let

$$q = \sqrt{mn} = \sqrt{\frac{dx}{t} \cdot \frac{dy}{t}} = \frac{d}{t}\sqrt{xy},$$

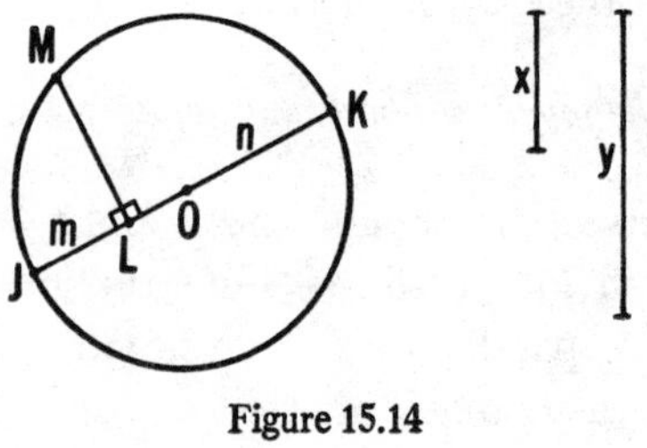

Figure 15.14

so the required $\sqrt{xy}$ is the fourth proportional to d, q, t. Thus we have reduced our problem to finding the mean proportional of the special segments m and n which add up to d.

To find q, take any diameter JOK of the fixed circle (see Figure 15.14). On JK mark off (by 3) $JL = m$; then $LK = d - m = n$. Construct (by 2) the perpendicular LM to JK. Then

$$LM = q = \sqrt{mn}$$

by an easy application of a well known theorem of elementary geometry, based on similar triangles.

Now we can easily handle the main construction problems (a) and (c). Again we take (c) first.

(c) *To find the points of intersection of a given circle* $X(r)$ *and straight line* AB.

The circle, of course, is not drawn; we are given merely its centre X, and its radius equal to segment r.

One related point we can find easily is T, the foot of the perpendicular from X to AB (see Figure 15.15). T, then, is the midpoint of the required chord MN. Also, this gives the length of the segment $XT = b$.

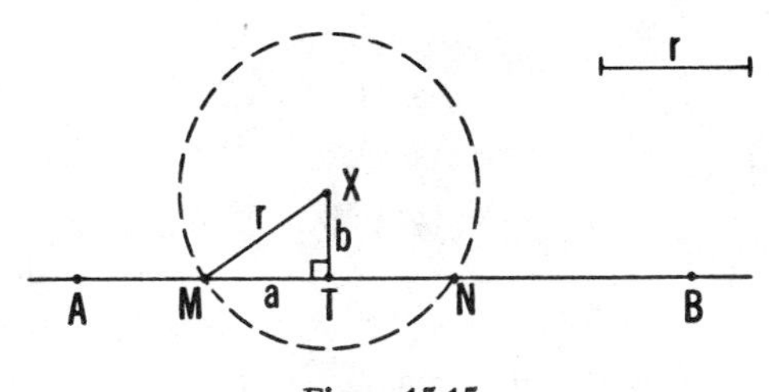

Figure 15.15

Now, in right triangle XMT (call $MT = a$),

$$a^2 = r^2 - b^2 = (r - b)(r + b).$$

Both r and b are known, so $r - b$ and $r + b$ can be determined by 3 above. By 5, then, we can determine the length of

$$a = \sqrt{(r - b)(r + b)},$$

the mean proportional of $r - b$ and $r + b$. By 3, we can mark off on AB the segment a on either side of T and obtain the required points of intersection M and N.

(a) *To find the intersections of two circles,* $X(r)$ *and* $Y(m)$.

Name the circles so that $Y(m)$ is at least as big as $X(r)$, and denote the distance XY between their centres by t. They will intersect in two points M and N if, for $t \geq m$, $t < m + r$ (see Figure 15.16), and for $t < m$, if $r > m - t$ (see Figure 15.17). The line through M and N is perpendicular to the line XY. Consequently, the law of cosines applied to $\triangle MXY$ yields

$$m^2 = r^2 + t^2 - 2tr \cos \measuredangle MXY.$$

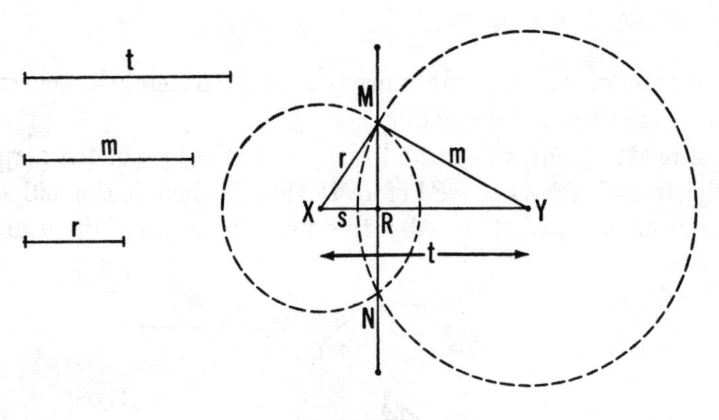

Figure 15.16

We denote XR by s and consider s positive if R is on XY (Figure 15.16) and negative if R is on XY extended (Figure 15.17). In all cases, $\cos \measuredangle MXY = s/r$, so that

$$m^2 = r^2 + t^2 - 2ts,$$

whence

$$s = \frac{r^2 + t^2 - m^2}{2t}.$$

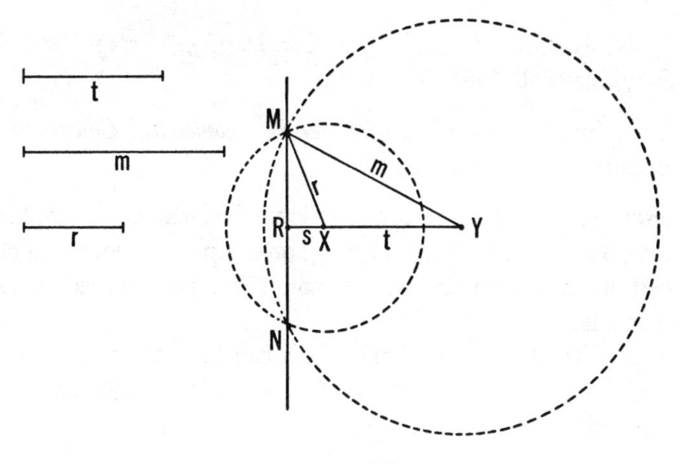

Figure 15.17

We shall construct

$$s = \frac{r^2}{2t} + \frac{t^2}{2t} - \frac{m^2}{2t}$$

from the information we have, namely r, t and m. These three terms are the fourth proportionals to

$$2t, r, r; \qquad 2t, t, t; \qquad \text{and} \qquad 2t, m, m,$$

respectively. Each can be constructed according to 4 (after $2t$ is found from t by 3), and the combination

$$\frac{r^2}{2t} + \frac{t^2}{2t} - \frac{m^2}{2t}$$

is then obtained by two applications of 3.

Thus we can mark R on XY (by 3) and construct the perpendicular PQ to XY at R. Our problem is now reduced to finding the intersection points of either of our given circles with the line PQ. But we just solved this problem above, in (c).

In closing this essay we note two excellent little books on these construction problems:

1. A. N. Kostovskii, *Geometrical Constructions Using Compasses Only*, Blaisdell, 1961, New York.
2. A. S. Smogorzhevskii, *The Ruler in Geometrical Constructions*, Blaisdell, 1961, New York.

1. discusses such things as constructions which can be accomplished with compasses set at a fixed opening; or compasses allowed to open to a certain maximum amount; or compasses not allowed to close below some limit.

2. deals with the constructions which can be made with straightedge and a square (instead of the fixed circle); straightedge and a pair of parallel lines; straightedge and an arc of a circle.

ADDITIONAL REFERENCES

H. Dorrie, 100 *Great Problems of Elementary Mathematics*, Dover, 1958, New York.

L. S. Shively, *An Introduction to Modern Geometry*, Wiley, 1939, New York.

ESSAY SIXTEEN

A Property of Some Repeating Decimals

Students learn early in school that a rational number n/d (a ratio of integers n, d) can be converted to a decimal. Moreover, if n/d is in lowest terms and all prime factors of the denominator d divide the base 10, then n/d has a terminating decimal expansion (see Exercise 1). In all other cases, n/d yields a repeating decimal. The block of digits which repeat is called a *period* of the unending decimal, and the number of digits in the shortest repeating block is called the *length* of the period. The usual way of seeing that a rational number n/d, where d has factors other than 2 and 5, has a periodic decimal expansion is to examine the process of long division. When n is divided by d, the only possible remainders are the numbers $1, 2, \cdots, d-1$. Therefore, after at most $d-1$ steps in the long division, one of these remainders must turn up again, and when this happens the long division process just repeats the previous steps and yields the same sequence of quotients and remainders as before (see Exercise 2).

In this section we shall investigate rational numbers with *prime* denominators whose decimal expansions have a period of even length;

some examples are

$$\text{(i)} \quad \tfrac{1}{7} = .142857142857\cdots = .\overline{142857},$$

$$\text{(ii)} \quad \tfrac{1}{13} = .\overline{076923},$$

$$\text{(iii)} \quad \tfrac{1}{17} = .\overline{0588235294117647},$$

$$\text{(iv)} \quad \tfrac{1}{19} = .\overline{052631578947368421}.$$

(A bar has been placed over the period in each.)

At first glance the blocks on the right do not appear specially interesting. But a closer look reveals that these examples share the remarkable property that digits which are half a period-length apart add up to 9. If one takes the shortest periods in these examples, divides them in half and adds the two parts, considering each as an ordinary number,

$$\text{(i) gives } \begin{array}{r} 142 \\ \underline{857} \\ 999; \end{array} \qquad \text{(iii) gives } \begin{array}{r} 05882352 \\ \underline{94117647} \\ 99999999; \end{array}$$

$$\text{(ii) gives } \begin{array}{r} 076 \\ \underline{923} \\ 999; \end{array} \qquad \text{(iv) gives } \begin{array}{r} 052631578 \\ \underline{947368421} \\ 999999999. \end{array}$$

Is there a general pattern behind these occurrences?

The answer is "yes": *If the repeating decimal expansion of the rational number n/p, where p is a prime not equal to 2 or 5, has a shortest period containing an even number of digits, then the sum of the first and second halves of that period, considered as ordinary numbers, consists entirely of* 9's.

Our main tools in proving the above assertion will be a bit of modular arithmetic leading to Fermat's (little) Theorem, and some familiarity with sums of geometric series. Since the reader is probably familiar with geometric progressions, let us recall these matters first.

The identity

$$(1 + b + b^2 + \cdots + b^n)(1 - b) = 1 - b^{n+1}$$

implies that the sum of the finite geometric series

$$a + ab + ab^2 + \cdots + ab^n = \frac{a(1 - b^{n+1})}{1 - b} \qquad \text{for } b \neq 1.$$

If $0 < b < 1$, and if n tends to infinity, then b^{n+1} tends to zero. Hence

$$a + ab + ab^2 + \cdots = \frac{a}{1 - b}. \tag{1}$$

We apply this result to repeating decimals, say to Example (i) on page 148, and find that

$$\frac{1}{7} = \frac{142857}{10^7} + \frac{142857}{10^{14}} + \frac{142857}{10^{21}} + \cdots$$

$$= \frac{142857}{10^7}\left(1 + \frac{1}{10^7} + \cdots\right) = \frac{142857}{10^7\left(1 - \frac{1}{10^7}\right)} = \frac{142857}{10^7 - 1}.$$

More generally, if a decimal has a period represented by the number P of u digits, and if this period starts after s places to the right of the decimal point, we may write the decimal in the form

$$\frac{n}{d} = R + \frac{Q}{10^s} + \frac{P}{10^{s+u}} + \frac{P}{10^{s+2u}} + \frac{P}{10^{s+3u}} + \cdots$$

$$= R + \frac{Q}{10^s} + \frac{1}{10^s}\left[\frac{P}{10^u} + \frac{P}{10^{2u}} + \cdots\right]$$

$$= R + \frac{Q}{10^s} + \frac{1}{10^s} \cdot \frac{P}{10^u - 1}.$$

If $n < d$, then the integer part R is zero; and if, moreover, the period of the decimal begins at the decimal point, then $s = 0$, and we have

$$(2) \qquad \frac{n}{d} = \frac{P}{10^u} + \frac{P}{10^{2u}} + \cdots = \frac{P}{10^u - 1}.$$

Next we shall develop the tools to show that *when p is a prime which does not divide* 10, *n/p is periodic, and its period begins at the decimal point.* It suffices to consider only proper fractions; for, if $n > p$, say $n = pq + r$ with $r < p$, then

$$\frac{n}{p} = q + \frac{r}{p},$$

the integer q precedes the decimal point, and the expansion of r/p follows it.

In dividing a number by p, the following non-zero remainders can occur:

$$1, 2, 3, \cdots, p - 1.$$

We shall say that two integers are *congruent modulo p* if their difference is divisible by p. Another way of saying this is that two numbers are congruent modulo p if they have the same remainder on division by p. The symbolic way of writing this is

$$f \equiv g \pmod{p} \qquad \text{(read “}f\text{ is congruent to }g\text{ mod }p\text{”)},$$

and it means that there is an integer c such that

$$f - g = cp, \qquad \text{so} \qquad f = cp + g.$$

It is easy to show [see Exercise 3(a)] that the set of all the integers may be partitioned into p equivalence classes so that each integer belongs to exactly one class. All integers divisible by p belong to the class corresponding to the remainder 0; all integers having the remainder 1 upon division by p belong to the class corresponding to 1; and so on for each possible remainder.

An essential property of congruence is that if $f_1, f_2, \cdots, f_m$ are congruent, respectively, to $g_1, g_2, \cdots, g_m$ modulo p, then the product of the f's is congruent to the product of the g's [see Exercise 3(b), (c)]; that is, if

$$(3) \qquad f_i \equiv g_i \pmod{p}, i = 1, 2, \cdots, m,$$

then

$$f_1 f_2 \cdots f_m \equiv g_1 g_2 \cdots g_m \pmod{p}.$$

Now consider any number a not divisible by the prime p. Form the numbers

$$a, 2a, 3a, \cdots, (p-1)a.$$

We claim that *no two of these* $p-1$ *numbers are congruent to each other modulo* p. To see this, suppose that

$$la \equiv ma \pmod{p}, \qquad 1 \leq l \leq p-1, \qquad 1 \leq m \leq p-1.$$

This means $la - ma$ is divisible by p, so there is an integer w such that

$$a(l-m) = wp.$$

Since the prime p divides the integer on the right, it must divide the integer on the left, and since p does not divide a, p must divide $l - m$. But l and m belong to the set $\{1, 2, \cdots, p-1\}$, so their difference is less in absolute value than p, hence cannot be divisible by p unless $l = m$. This proves our claim.

We now consider the products

$$\Pi_1 = 1 \cdot 2 \cdot 3 \cdots (p-1)$$

and

$$\Pi_2 = a \cdot 2a \cdot 3a \cdots (p-1)a = a^{p-1}\Pi_1.$$

Since each of the $p-1$ factors in Π_1 is congruent to precisely one of the $p-1$ factors of Π_2 we conclude by means of the assertion

(3) [to be proved in Exercise 3(c)] that

$$\Pi_1 \equiv a^{p-1}\Pi_1 \pmod{p};$$

that is,

$$a^{p-1}\Pi_1 - \Pi_1 = \Pi_1(a^{p-1} - 1)$$

is divisible by p. Observe that Π_1 is not divisible by p (none of its factors is), so $a^{p-1} - 1$ is divisible by p.

We restate the theorem we have just proved.

FERMAT'S THEOREM. *If p is a prime and a any number not divisible by p, then*

$$a^{p-1} \equiv 1 \pmod{p}, \tag{4}$$

that is, $a^{p-1} - 1$ is divisible by p.

We now set $a = 10$. If p is a prime other than 2 or 5, Fermat's theorem guarantees that there is an integer c such that

$$10^{p-1} - 1 = cp.$$

Taking reciprocals of both sides and then multiplying by c yields

$$\frac{1}{p} = \frac{c}{10^{p-1} - 1},$$

so that

$$\frac{r}{p} = \frac{rc}{10^{p-1} - 1}. \tag{5}$$

Since $r < p$, we know that $rc < 10^{p-1}$, hence rc has at most $p - 1$ digits. We recognize that the right member of (5) is the sum of the geometric series

$$\frac{r}{p} = \frac{rc}{10^{p-1}} + \frac{rc}{10^{2(p-1)}} + \cdots$$

which represents a repeating decimal with the repeating block of $p-1$ digits represented by the number rc. Thus *the length of the shortest period of* r/p *is either* $p-1$ *or some divisor of* $p-1$. [Observe that in Examples (i), (iii), (iv), the shortest periods have lengths $p-1$, while in Example (ii), the period has length $(p-1)/2$.]

The same reasoning that deduced (5) from Fermat's Theorem yields:

If k *is any positive integer such that* p *divides* $10^k - 1$, *then*

$$\frac{r}{p} = \frac{R}{10^k - 1}, \tag{6}$$

where R *is a positive integer less than* 10^k *so that the expansion of* r/p *repeats after* k *digits.*

Conversely, *if the expansion of* r/p *repeats after* k *digits, then* [cf. equation (2), p. 150].

$$\frac{r}{p} = \frac{P}{10^k - 1} \quad \text{(where } P \text{ is the period of the expansion of } r/p\text{)},$$

so

$$pP = r(10^k - 1),$$

p does not divide r, and hence p *divides* $10^k - 1$.

We conclude that the decimal expansion of r/p repeats after k digits if and only if $10^k - 1$ is divisible by p. In particular, *the number of digits* u *in the shortest period of* r/p *is the smallest integer* k *such that* $10^k - 1$ *is divisible by* p.

Observe that the length of the period of r/p depends only on p, not on r.

We shall now prove the assertion on page 148 illustrated by Examples (i)–(iv). Here we assume that the shortest period P of the decimal expansion of r/p, p a prime not equal to 2 or 5, has an even number, say $2t$, of digits. We write P in the form

$$P = M \cdot 10^t + N, \qquad N < 10^t, \tag{7}$$

where M and N are t-digit numbers making up the two halves of P; we wish to show that $M + N = \underbrace{99\cdots9}_{t \text{ digits}} = 10^t - 1.$

According to (2), page 150,

$$\text{(8)} \qquad \frac{r}{p} = \frac{P}{10^{2t} - 1},$$

so that

$$pP = r(10^{2t} - 1) = r(10^t - 1)(10^t + 1).$$

Since r/p is a reduced fraction, p does not divide r. Moreover, p does not divide $10^t - 1$; for, if it did, the length of the period of r/p would be t rather than $2t$ (see page 153). Hence the prime p divides $10^t + 1$. We use the representation (7) for P to write (8) in the form

$$\frac{r}{p} = \frac{P}{10^{2t} - 1} = \frac{M \cdot 10^t + N}{(10^t - 1)(10^t + 1)},$$

whence

$$\text{(9)} \qquad \begin{aligned} \frac{r(10^t + 1)}{p} &= \frac{M \cdot 10^t + N}{10^t - 1} = \frac{M \cdot 10^t - M + M + N}{10^t - 1} \\ &= \frac{M(10^t - 1) + M + N}{10^t - 1} = M + \frac{M + N}{10^t - 1}. \end{aligned}$$

The first member of equations (9) is a positive integer because p divides $10^t + 1$, so the last member is an integer.
The numbers M and N have t digits each, that is,

$$0 < M + N \leq 10^t - 1 + 10^t - 1,$$

so that

$$0 < \frac{M + N}{10^t - 1} \leq 2.$$

Equality could hold only if $M = N = 10^t - 1$, in which case each digit in P would be 9; but then the shortest period would have one digit contrary to the assumption that the length of the (shortest) period is even. Therefore

$$\frac{M+N}{10^t - 1} = 1,$$

so that

$$M + N = 10^t - 1$$

as we set out to prove.

EXERCISES

1. (a) Let n, d be positive integers, and suppose the prime factorization of d is $d = 2^k 5^l$ where k, l are non-negative integers. Show that the decimal expansion of n/d terminates. [*Hint*: A terminating decimal is of the form $D/10^m$.]

   (b) State and prove the analogous assertion for the expansion of n/d in a system with base b.

2. By analyzing the long division algorithm, prove that a rational number n/d with a non-terminating decimal expansion has, in fact, a periodic expansion. (*Hint*: $n = q_1 d + r_1$, $10r_1 = q_2 d + r_2$, $10r_2 = q_3 d + r_3$, $\cdots$ are the steps figuring in long division, with $r_1, r_2, \cdots$ successive remainders.)

3. (a) A relation $\sim$ is called an "equivalence relation" if it is

   1) reflexive ($f \sim f$ for all elements f),
   2) symmetric (if $f \sim g$ then $g \sim f$) and
   3) transitive (if $f \sim g$ and $g \sim h$ then $f \sim h$).

   Using the definition of congruence, show that $\equiv \pmod{p}$ is an equivalence relation.

   (b) Prove that if $f_1 \equiv g_1 \pmod{p}$ and $f_2 \equiv g_2 \pmod{p}$, then $f_1 f_2 \equiv g_1 g_2 \pmod{p}$. (*Hint*: By definition, $f_1 - g_1 = k_1 p$, $f_2 - g_2 = k_2 p$; multiply the first relation by g_2, the second by f_1, and add them.)

(c) Using the result of part (b), prove: if $f_i \equiv g_i \pmod{p}$ for $i = 1, 2, \cdots, m$, then $f_1 f_2 \cdots f_m \equiv g_1 g_2 \cdots g_m \pmod{p}$.

4. Consider a numeration system with base b. State and prove the analogue of the assertion concerning the expansion of a rational number with prime denominator whose period is of even length.

5. Let q be any number relatively prime to 10. Show that the decimal expansion of r/q is periodic, with period beginning at the decimal point.

6. Prove the following generalization of Fermat's theorem: Let q be any number relatively prime to a. Let $\varphi(q)$ be the number of integers in the set $1, 2, \cdots, q-1$ which are relatively prime to q. Then

$$a^{\varphi(q)} \equiv 1 \pmod{p}.$$

REFERENCES

H. Rademacher and O. Toeplitz, *The Enjoyment of Mathematics*, Princeton University Press, 1957, Princeton.

G. Hardy and E. Wright, *Introduction to the Theory of Numbers*, Clarendon, 1960, Oxford.

H. N. Wright, *First Course in Theory of Numbers*, Wiley, 1939, New York.

ESSAY SEVENTEEN

The Theorem of Barbier

The width of a closed curve in one direction may be different from its width in another direction. Taking two lines, t and t', perpendicular to a given line AB and far enough apart so that a given closed curve Q lies between them, we determine the width of Q in the direction of AB by bringing together t and t' until contact is made with Q (Figure 17.1). The required width is then taken to be the distance between the final positions of t and t'. (We note that t and t' are always parallel.) Such lines as t and t' (in their final positions) which make contact with Q and have Q entirely on one side of them are called *supporting lines* for Q.

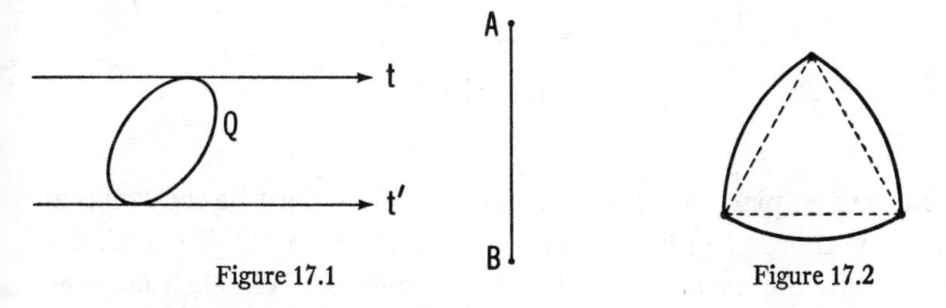

Figure 17.1

Figure 17.2

If the width of Q is the same for every possible direction AB, Q is said to be a curve of *constant width.* Obviously, a circle is such a curve. However, one may be momentarily at a loss to imagine any other shape which has constant width. A so-called Reuleaux triangle, illustrated in Figure 17.2, is one example of many such figures. It is constructed by drawing arcs with centres at the vertices A, B, C of an equilateral triangle ABC and radius equal to a side of the triangle.

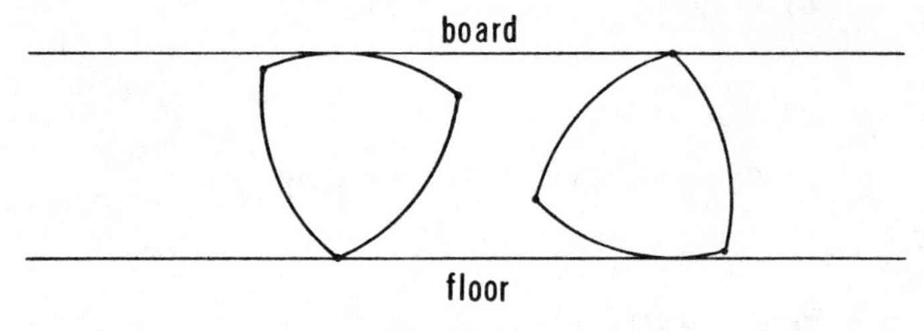

Figure 17.3

A board put on top of a pair of such figures of the same size so that they act as rollers, will receive a perfectly smooth ride as the figures turn (Figure 17.3). While there are a great many different curves of constant width (some "lopsided", as in Figure 17.4), they all have certain properties in common, which we now consider.

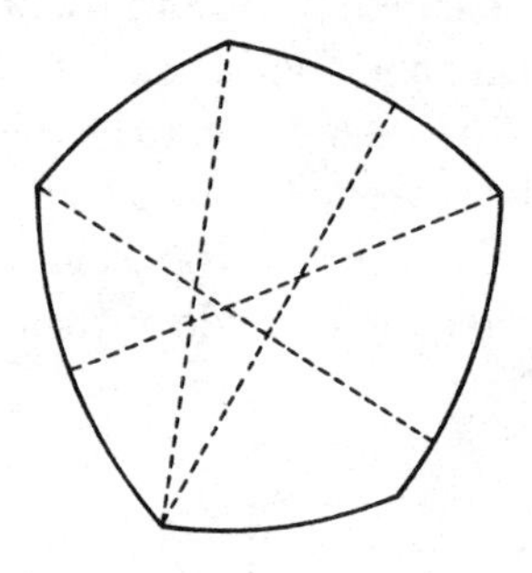

Figure 17.4

In the first place, a curve Q of constant width must be convex; that is, every chord of Q lies in the region bounded by Q. We shall not prove this, but rather look at things conversely. We shall consider those convex curves which do have constant width. (That this constitutes all curves of constant width is immaterial to us.) One of the most striking properties of such curves is given by the Theorem of Barbier:

All convex curves of constant width b have the same perimeter.

Since the circle of diameter b is obviously among such curves, the common perimeter is πb. It is to the proof of this remarkable theorem that we devote this essay.

We first establish a few preliminary results needed later. Throughout we denote by Q a curve of constant width and by b its uniform width. Lemma 1 is obvious.

LEMMA 1. *If two parallel lines both intersect Q, their distance apart is not greater than b.*

LEMMA 2. *If A and B lie on Q, then $AB \leq b$.*

Proof: Draw t and t' through A and B respectively, each perpendicular to AB. Then AB is the distance between a pair of parallel lines which intersect Q. By Lemma 1, it follows that $AB \leq b$.

LEMMA 3. *If parallel supporting lines t and t' have points A and B, respectively, in common with Q, then AB is perpendicular to both t and t'.*

Proof: By Lemma 2, we have $AB \leq b$. Because t and t' are parallel supporting lines, their distance apart is by definition equal to the width b. The distance from a point (A) on one of them to a point (B) on the other is at least as big as their distance apart; that is, $AB \geq b$. Hence $AB = b$. It follows, then, that AB is perpendicular to both t and t'.

LEMMA 4. *Parallel supporting lines t and t' each meet Q in only one point.*

Proof: Suppose, if possible, that two points, A and C, both are on t and on Q. Then, letting B be a point of contact of Q with t', we have by Lemma 3 that both AB and CB are perpendicular to t and t'. This means that two perpendiculars can be drawn from B to t—a contradiction. Hence Q meets t in only one point. (Similarly for t'.)

Now let us proceed to the proof of Barbier's theorem.

Proof of Barbier's theorem: We are to show that the perimeter of a given convex curve Q of constant width b is πb. As might be expected, we do this by comparing the perimeter of Q with the circumference of a circle of diameter b.

Around a circle of diameter b construct a regular polygon of n sides, P_n. From Figure 17.5, we see that

$$\measuredangle UOW = 2\pi/n, \qquad \text{and} \qquad \measuredangle UOV = \pi/n; \qquad OV = b/2.$$

Letting s represent the length of each side of P_n, we obtain from $\triangle UOV$ that

$$\frac{s/2}{b/2} = \tan \pi/n, \qquad \text{so} \qquad s = b \tan \pi/n.$$

The perimeter of P_n is n times s:

$$\text{perimeter of } P_n = nb \tan \pi/n.$$

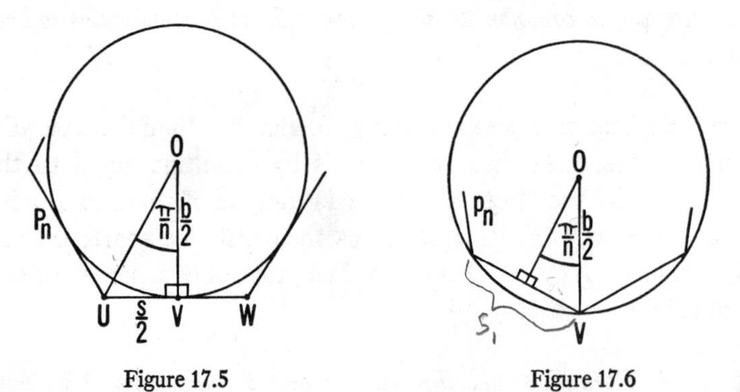

Figure 17.5 Figure 17.6

Now P_n has n points of contact with the circle, points such as V in Figure 17.5, which may be taken as the n vertices of a regular polygon p_n inscribed in the circle. From Figure 17.6, we see that

$$\text{perimeter of } p_n = nb \sin \pi/n.$$

Clearly the perimeter of p_n is less than πb (the perimeter of the circle) while that of P_n is greater than πb; that is,

$$nb \sin \pi/n < \pi b < nb \tan \pi/n.$$

As n increases, the perimeter of P_n diminishes toward the circumference of the circle, and the perimeter of p_n increases toward it. Each of these perimeters approaches the circumference of the circle as the number of sides n approaches infinity. In fact, the common limit of these perimeters is the very definition of the length of the perimeter of the circle.

Now then, settle on some value of n and compute the perimeters of P_n and p_n. Next, construct around the circle a regular polygon of $2n$ sides, P_{2n}; since P_{2n} has an even number of sides, pairs of opposite sides of this polygon are parallel. Proceeding to the curve Q, draw n pairs of parallel supporting lines in the same directions, respectively, as the n pairs of opposite parallel sides of P_{2n}. This gives us a polygon Q_{2n} of $2n$ sides circumscribed about Q. By Lemma 4, each of its sides meets Q in only one point. The $2n$ points thus obtained provide an inscribed polygon of $2n$ sides, q_{2n}. We note that Q_{2n} and q_{2n} need not be regular polygons.

The length of the perimeter of Q (by definition) is the common limit of the perimeters of Q_{2n} and q_{2n} as n approaches infinity. We must show, then, that both the perimeter of Q_{2n} and the perimeter of q_{2n} have the limit πb.

What we do know is that the perimeters of P_n and p_n converge to πb as n tends to infinity. Consequently, if we could show that the perimeters of Q_{2n} and q_{2n}, for every value of n, lie *between* the perimeters of P_n and p_n, it would follow that the perimeters of Q_{2n} and q_{2n} also converge to πb. Our task, then, is to prove that

$$\text{perimeter of } p_n < \text{perimeter of } q_{2n} < \text{perimeter of } Q_{2n} < \text{perimeter of } P_n$$

(see Figure 17.7).

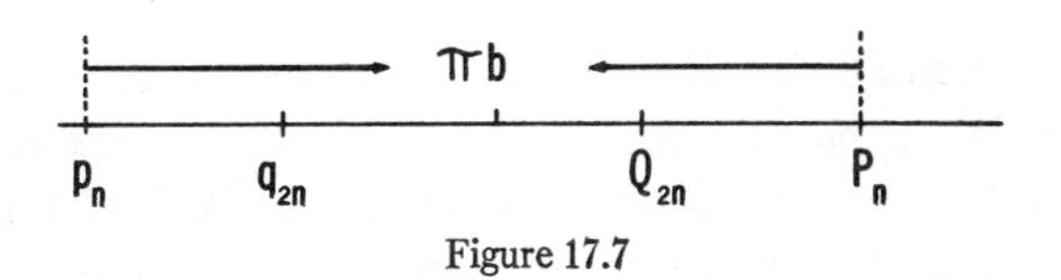

Figure 17.7

Let the vertices of Q_{2n} be $A_1, A_2, \cdots, A_{2n}$, and those of q_{2n} be $B_1, B_2, \cdots, B_{2n}$. The vertex with subscript $i+n$ is opposite that with subscript i, since the $2n$ vertices are consecutively ordered.

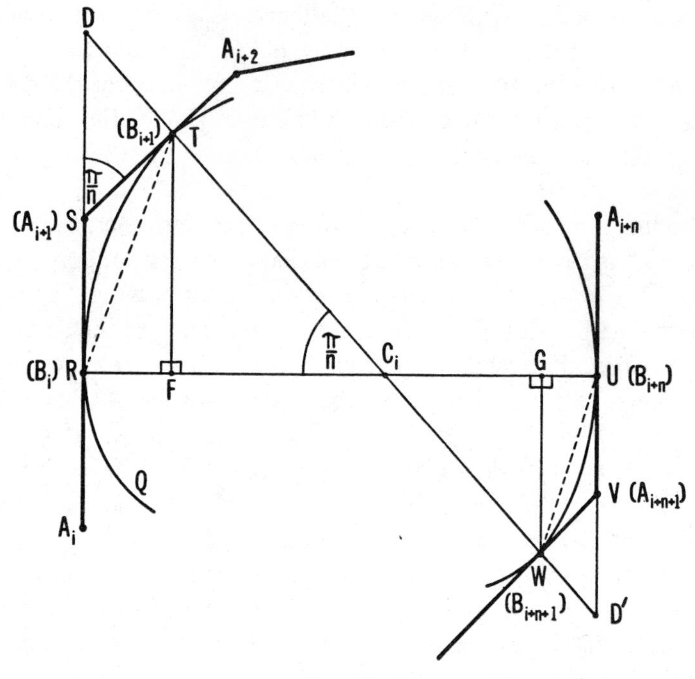

Figure 17.8

Also, let B_i lie on the side A_iA_{i+1} of Q_{2n} (Figure 17.8). We want to consider the two opposite parts of the figure in the neighbourhoods of B_iB_{i+1} and $B_{i+n}B_{i+1+n}$. The subscripts make reading difficult; hence, let us re-name the six important points with which we are concerned as follows:

$$B_i = R; \qquad A_{i+1} = S; \qquad B_{i+1} = T;$$
$$B_{i+n} = U; \qquad A_{i+1+n} = V; \qquad B_{i+1+n} = W.$$

We note first of all that R and U are points of contact of Q with a pair of parallel supporting lines (sides of Q_{2n}). Consequently, by Lemma 3, the segment RU is perpendicular to both of these supporting lines. Similarly for the line segment TW joining the next pair of opposite vertices of q_{2n}. Now produce the line TW to meet the sides A_iS and $A_{i+n}V$ of Q_{2n}, extended, in D and D', respectively, as shown in Figure 17.8. Denote by C_i the point where TW intersects RU. Because the angles at R and T are right angles, the quadrilateral $RSTC_i$ is cyclic. This means that the exterior angle at S is the same as the interior angle at C_i. Now the sides of Q_{2n} are parallel to the sides of a *regular* polygon P_{2n}. Consequently, the amount of turning from one side to the next, i.e., the size of an exterior angle of P_{2n} or of Q_{2n}, is the same at all vertices, namely $2\pi/2n$ or π/n. Thus the angle at C_i is π/n.

Now SD, being the hypotenuse of a right triangle, is greater than the side ST. Consequently,

$$RS + ST < RS + SD = RD.$$

Because $\triangle RDC_i$ is a right triangle, $RD = RC_i \tan \pi/n$. Thus

$$RS + ST < RC_i \tan \pi/n. \tag{1}$$

Similarly, at the opposite corner,

$$UV + VW < UV + VD' = UD' = UC_i \tan \pi/n. \tag{2}$$

Inequalities (1) and (2) imply that

$$\begin{aligned} RS + ST + UV + VW &< (RC_i + UC_i) \tan \pi/n \\ &= RU \tan \pi/n = b \tan \pi/n, \end{aligned}$$

since the width RU of Q is b. The quantity $b \tan \pi/n$ is independent of the pair of opposite vertices R and U that are chosen.

At each pair A_{i+1}, A_{i+n+1} of opposite vertices of Q_{2n} we add those pieces of adjacent sides which lie between the vertices B_i, B_{i+1} and B_{i+n}, B_{i+n+1}, as we did at S and V. After summing these pieces at each of the n pairs of opposite vertices of Q_{2n}, we have taken account of the entire perimeter. Our n additions yield

$$\text{perimeter of } Q_{2n} < nb \tan \pi/n,$$

and, since the perimeter of P_n is $nb \tan \pi/n$ (see page 160), we have shown that

$$\text{perimeter of } Q_{2n} < \text{perimeter of } P_n.$$

To show that perimeter of q_{2n} > perimeter of p_n, drop perpendiculars TF and WG to RU (see Figure 17.8). Then

$$RT > TF = TC_i \sin \pi/n, \qquad UW > WG = C_iW \sin \pi/n,$$

so

$$RT + UW > (TC_i + C_iW) \sin \pi/n = TW \sin \pi/n = b \sin \pi/n.$$

The sum of each pair of opposite sides of q_{2n} satisfies such an inequality, and since there are n pairs of opposite sides, we have

$$\text{perimeter of } q_{2n} > nb \sin \pi/n.$$

The right side is the perimeter of p_n (see page 160). This completes the proof of Barbier's theorem.

NOTE: Our considerations are valid even if the points C_i, R, S, T all coincide, as they may well do, for example, at a vertex of a Reuleaux triangle.

REFERENCES

L. Lyusternik, *Convex Figures and Polyhedra*, Dover, 1963, New York.

I. M. Yaglom and V. G. Boltyanski, *Convex Figures*, Holt, Rinehart and Winston, 1961, New York.

ESSAY EIGHTEEN

The Series of Reciprocals of the Primes

In this section, we consider the series

$$(P) \qquad \frac{1}{2}+\frac{1}{3}+\frac{1}{5}+\cdots+\frac{1}{p}+\cdots,$$

where the denominators p are the prime numbers in ascending order. We will refer to this as series P. It is an especially interesting series to deal with.

We begin with a few basic notions concerning series of positive numbers. We emphasize that our remarks apply to series of *positive* terms; considering just these series we can get away with a minimum of background. The treatment of series with both positive and negative terms is considerably more demanding. We define three series A, B, C as follows:

$$(A) \qquad 1+2+3+4+\cdots+n+\cdots;$$

$$(B) \qquad 1+\frac{1}{2}+\frac{1}{4}+\frac{1}{8}+\cdots+\frac{1}{2^{n-1}}+\cdots;$$

$$(C) \qquad 1+\frac{1}{2}+\frac{1}{3}+\frac{1}{4}+\cdots+\frac{1}{n}+\cdots.$$

The fundamental question about series concerns the "sum" of all the terms. This terminology is explained as follows.

By the n-th *partial sum* of an infinite series, we mean the sum of the first n terms. For example, the partial sums $A_1, A_2, \cdots, A_n$ of series A above are

$$A_1 = 1, \quad A_2 = 1 + 2 = 3, \quad A_3 = 1 + 2 + 3 = 6, \cdots,$$
$$A_n = 1 + 2 + \cdots + n = \tfrac{1}{2}n(n+1), \cdots.$$

Since we are considering only series with positive terms, the partial sums *increase steadily* with the inclusion of each term. One of two things then must happen: either the partial sums become arbitrarily large, or there is some number which they never exceed. For the series A above, the sums increase beyond all bounds. The quantity $\frac{1}{2}n(n+1)$ grows without limit as n increases. Such series are called *divergent* and are said *not* to have a sum. The partial sums $B_1, B_2, \cdots$ for series B, however, are bounded; they are

$$B_1 = 1, \quad B_2 = 1 + \tfrac{1}{2} = \tfrac{3}{2}, \quad B_3 = 1 + \tfrac{1}{2} + \tfrac{1}{4} = \tfrac{7}{4}, \quad B_4 = \tfrac{15}{8}, \cdots,$$

and, according to the formula for the sum of the first n terms of a geometric progression,

$$B_n = 1 + \tfrac{1}{2} + \cdots + (\tfrac{1}{2})^{n-1} = \frac{1 - (\frac{1}{2})^n}{1 - \frac{1}{2}} = 2 - \frac{1}{2^{n-1}}.$$

Clearly, no matter how big n is, the partial sum B_n is less than 2.

Now, if the partial sums of a series are bounded by a number M, they are also bounded by every number larger than M (e.g., series B is not only bounded by 2, but by every number greater than 2). It may happen, however, that there is no bound which is smaller than M. (Series B, for example, surpasses *any* number which is less than 2.) In the case of bounded series, or *convergent* series as they are called, the *smallest bound* S is said to be the *sum* of the series.† (The sum of series B is 2).

† This is often stated as follows: An infinite series converges to a number S (called its sum) if the sequence $S_1, S_2, S_3, \cdots$ of its partial sums has the limit S, that is, if $\lim_{n \to 0} S_n = S$.

Before launching into the main topic of series P, let us digress briefly with series C above. C provides an unexpected result: even though the terms approach zero, the series diverges. We see this by dividing the series into sections, each of which ends with a term of the form $1/2^n$, $n = 1, 2, \cdots$.

In the first group we put the terms up to $\dfrac{1}{2^1}$: $1 + \dfrac{1}{2}$;

in the second group go the succeeding terms up to $\dfrac{1}{2^2}$: $\dfrac{1}{3} + \dfrac{1}{4}$;

in the third group go the next terms up to $\dfrac{1}{2^3}$: $\dfrac{1}{5} + \cdots + \dfrac{1}{8}$;

....

the n-th section is $\dfrac{1}{2^{n-1}+1} + \dfrac{1}{2^{n-1}+2} + \cdots + \dfrac{1}{2^n}$, etc.

Since the terms of the series steadily decrease, the last term in each section is the smallest. Thus the sum of the terms of each section is bigger than what would be obtained if each term in the section were replaced by the last one. But this smaller sum is equal to $\frac{1}{2}$ for every $n > 1$. Thus

$$1 + \tfrac{1}{2} > \tfrac{1}{2} + \tfrac{1}{2} = 1$$
$$\tfrac{1}{3} + \tfrac{1}{4} > \tfrac{1}{4} + \tfrac{1}{4} = \tfrac{1}{2}$$
$$\tfrac{1}{5} + \tfrac{1}{6} + \tfrac{1}{7} + \tfrac{1}{8} > \tfrac{1}{8} + \tfrac{1}{8} + \tfrac{1}{8} + \tfrac{1}{8} = \tfrac{1}{2}$$

....

$$\frac{1}{2^{n-1}+1} + \frac{1}{2^{n-1}+2} + \cdots + \frac{1}{2^n} > \underbrace{\frac{1}{2^n} + \frac{1}{2^n} \cdots + \frac{1}{2^n}}_{2^n - 2^{n-1} = 2^{n-1} \text{ terms}} = \frac{1}{2}$$

....

Since there is no end to the number of sections, there is no end to the number of $\frac{1}{2}$'s which go into the accumulated sum. And by adding enough $\frac{1}{2}$'s one can surpass any amount.

We see, then, that series B converges, while C diverges. We note that series B is what is left of C after deleting some of its terms. In fact, enough terms of C are removed to bring down the sum to a finite number. Now the series P of reciprocals of the primes can also be obtained from C by removing terms. The question arises whether enough terms are deleted to give a convergent series. Our main purpose in this essay is to give a neat proof that series P diverges.

We proceed by the indirect method; i.e., assuming that the series converges, we argue to a contradiction.

If series P converged to a sum S, then every number smaller than S would eventually be surpassed by the partial sums. In particular, the partial sums eventually exceed $S - \frac{1}{2}$. Suppose that this occurs upon the addition of the n-th term $1/p_n$ of the series:

$$\frac{1}{2} + \frac{1}{3} + \frac{1}{5} + \cdots + \frac{1}{p_{n-1}} \leq S - \frac{1}{2},$$

$$\frac{1}{2} + \frac{1}{3} + \frac{1}{5} + \cdots + \frac{1}{p_{n-1}} + \frac{1}{p_n} > S - \frac{1}{2}.$$

The sum of all the *remaining* terms,

$$\frac{1}{p_{n+1}} + \frac{1}{p_{n+2}} + \cdots,$$

then, must be less than $\frac{1}{2}$; for otherwise the grand total would be more than S, which is not the case. For *some* value of n, then, we have

$$\frac{1}{p_{n+1}} + \frac{1}{p_{n+2}} + \cdots < \frac{1}{2}.$$

We digress now into a seemingly unrelated area. Consider the sequence of prime numbers, in ascending order, beyond the k-th prime:

$$p_{k+1}, \quad p_{k+2}, \quad \cdots.$$

Let $N(x)$ represent the number of positive integers less than or equal to the positive integer x which are *not* divisible by any of the primes

$$p_{k+1}, \quad p_{k+2}, \quad \cdots.$$

In other words, $N(x)$ is the number of positive integers $\leq x$, all of whose prime factors are among the first k prime numbers $p_1, p_2, \cdots, p_k$. For instance, if $k = 4$, then

$$\{p_5, p_6, p_7, p_8, \cdots\} = \{11, 13, 17, 19, \cdots\},$$

and

$$N(10) = 10, \qquad N(15) = 13, \qquad N(27) = 20.$$

That is, no integers from 1 to 10 are divisible by $11, 13, 17, \cdots$, hence $N(10) = 10$; in the range from 1 to 15 only two integers (namely 11 and 13) are divisible by any of the primes $11, 13, 17, \cdots$, hence $N(15) = 13$; in the range 1 to 27, seven integers, 11, 13, 17, 19, 22, 23, 26, are divisible by one of the primes $11, 13, 17, \cdots$, hence $N(27) = 20$ for $k = 4$.

Let us suppose that some value for k has been decided upon and the values of $N(x)$ determined accordingly. (The particular value of k is immaterial at this point.) We can estimate the size of $N(x)$ as follows.

Let y represent any positive integer; it can be factored into its prime factors

$$y = p_1^{a_1} p_2^{a_2} p_3^{a_3} \cdots p_t^{a_t},$$

where some of the exponents $a_1, a_2, \cdots, a_t$ may be odd and some even. If an exponent a_i is odd, we decrease it to the next lower even value by transferring one of the factors p_i to the end of the expression. For example, if $y = 2^7 \cdot 3^4 \cdot 5^3 \cdot 7^6$, we rewrite it as

$$y = (2^6 \cdot 3^4 \cdot 5^2 \cdot 7^6)(2 \cdot 5).$$

In this way we may write $y = uv$, where every prime factor in u occurs an even number of times and every one in v occurs exactly once. Since the factors of u occur an even number of times, u is a perfect square; hence we may write

$$y = w^2 \cdot v,$$

where v possesses no repeated factors. Now, in order that an integer

y be counted by $N(x)$, it must fulfill the conditions

(i) $y \leq x$,

(ii) all prime factors of y must be among the first k primes $p_1, p_2, \cdots, p_k$.

In order to estimate $N(x)$, we estimate the number of positive integers $y = w^2v$ satisfying (i) and (ii). To satisfy (i), we must certainly stipulate that $w^2v \leq x$, and since $v \geq 1$, we must require that

$$w \leq \sqrt{x}.$$

To satisfy (ii) we must be sure that all prime factors of $y = w^2v$, and hence certainly all prime factors of v are among the first k primes $p_1, p_2, \cdots, p_k$. So we require v to be of the form

$$v = p_1^{b_1}p_2^{b_2} \cdots p_k^{b_k},$$

where, since v contains no repeated factors, each of the exponents $b_1, b_2, \cdots, b_k$ is either 0 or 1. Since only two choices are permissible for each of the k exponents, the total number of permissible values of v is at most 2^k. [It is less than 2^k if $p_1p_2 \cdots p_k > x$, because in this case, some values of v would violate condition (i).]

Now the number of values that $y = w^2v$ can take without violating conditions (i) and (ii) depends on the number of permissible values for its parts w and v. The number of choices for the parts are, respectively, at most $\sqrt{x}$ and at most 2^k. Consequently, there can be at most

$$2^k\sqrt{x}$$

suitable values for y; that is to say,

$$N(x) \leq 2^k\sqrt{x}.$$

There are x positive integers less than or equal to the integer x; $N(x)$ of these are not divisible by any of the primes $p_{k+1}, p_{k+2}, \cdots$, so the remaining $x - N(x)$ of them are divisible by some prime $p_{k+1}, p_{k+2}, \cdots$. We can try to count these $x - N(x)$ numbers directly by recording how many numbers between 1 and x are multiples of one of the primes $p_{k+1}, p_{k+2}, \cdots$. In the set $\{1, 2, \cdots, x - 1, x\}$, there are at most x/p_{k+1} multiples of p_{k+1} (since these multiples $p_{k+1}, 2p_{k+1}, 3p_{k+1}, \cdots$ occur only in every p_{k+1}-th position). For example, the set $\{1, 2, 3, \cdots, 19, 20\}$ contains the six multiples of 3: 3, 6, 9, 12, 15, 18; and $6 \leq \frac{20}{3}$. Similarly, at most x/p_{k+2} of the numbers from 1 to x are multiples of p_{k+2}, and so on. Therefore

$$x - N(x) \leq \frac{x}{p_{k+1}} + \frac{x}{p_{k+2}} + \cdots. \tag{1}$$

[We note that this estimate is definitely an over-estimate for large values of x; for example, when x is divisible by both p_{k+1} and p_{k+2}, it gets counted once in each of the expressions x/p_{k+1} and x/p_{k+2} although it contributes only one to $x - N(x)$.]

Now to get back to the main question. We have seen already that if the series P converges, then for some value of n we have

$$\frac{1}{p_{n+1}} + \frac{1}{p_{n+2}} + \cdots < \frac{1}{2} \tag{2}$$

(see p. 168).

Let us take k to be this value of n (whatever it is) and consider the corresponding functions $N(x)$ and $x - N(x)$. Relation (1) above, for $k = n$, becomes

$$x - N(x) \leq \frac{x}{p_{n+1}} + \frac{x}{p_{n+2}} + \cdots,$$

and multiplying relation (2) by x, we obtain (for any x)

$$\frac{x}{p_{n+1}} + \frac{x}{p_{n+2}} + \cdots < \tfrac{1}{2}x.$$

Combining these results, we get

$$x - N(x) < \tfrac{1}{2}x$$

which is equivalent to

$$\tfrac{1}{2}x < N(x)$$

and holds for *all values of* x. But earlier (see p. 170) we had that $N(x) \leq 2^k\sqrt{x}$, which for $k = n$ becomes

$$N(x) \leq 2^n\sqrt{x}.$$

Therefore

$$\tfrac{1}{2}x < N(x) \leq 2^n\sqrt{x}, \qquad \text{for all } x.$$

For x equal to 2^{2n+2}, we get, then,

$$\tfrac{1}{2}\cdot 2^{2n+2} < N(x) \leq 2^n\cdot 2^{n+1}, \qquad (\sqrt{x} = 2^{n+1}),$$

which simplifies to

$$2^{2n+1} < N(x) \leq 2^{2n+1}.$$

This contradiction proves that the series P diverges. (A similar contradiction is obtained for x equal to any value greater than 2^{2n+2}.)

REFERENCE

G. Hardy and E. Wright, *Introduction to the Theory of Numbers*, Clarendon, 1960, Oxford.

ESSAY NINETEEN

Van Schooten's Problem

The following problem was solved by the Dutch mathematician Franciscus Van Schooten (called Van Schooten the younger) (1615–1660): If a triangle ABC is moved in the plane so that A and B travel along straight lines m and n respectively, what is the locus of C?

His solution is very ingenious. First, we establish a preliminary result: *If a straight line l is moved so that two points on it, A and B, travel along two fixed, perpendicular straight lines, respectively, the locus of any third point on l is an ellipse.*

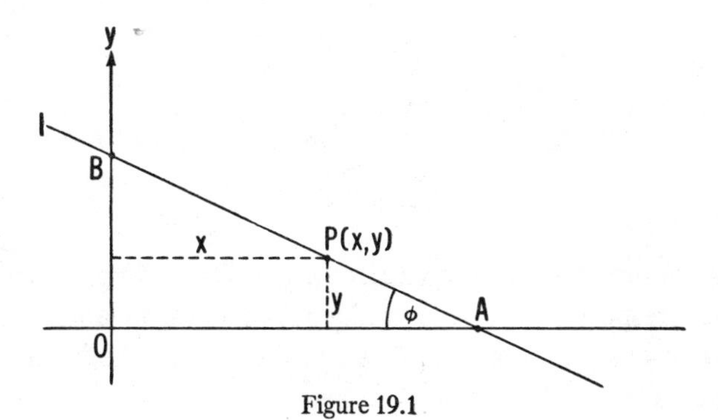

Figure 19.1

We take the fixed perpendicular lines as the x and y axes; as l moves, the point A is always on the x-axis, the point B on the y-axis. We denote the third point on the moving line by $P(x, y)$ and the changing angle OAB by ϕ (see Figure 19.1). If P is between A and B, then

$$(1) \qquad \frac{x}{BP} = \cos\phi, \qquad \frac{y}{AP} = \sin\phi.$$

If P is beyond either A or B, the expressions (1) still give the correct magnitudes for the coordinates x and y, but their signs may need changing. In any event, squaring formulas (1) gives results which hold for all positions of P; that is,

$$\text{(2)} \qquad \frac{x^2}{BP^2} = \cos^2 \phi, \qquad \frac{y^2}{AP^2} = \sin^2 \phi.$$

Consequently, we obtain

$$\cos^2 \phi + \sin^2 \phi = \frac{x^2}{BP^2} + \frac{y^2}{AP^2} = 1,$$

which tells us that the locus of P is an ellipse with centre at the origin and semi-axes BP and AP along the x and y axes respectively.

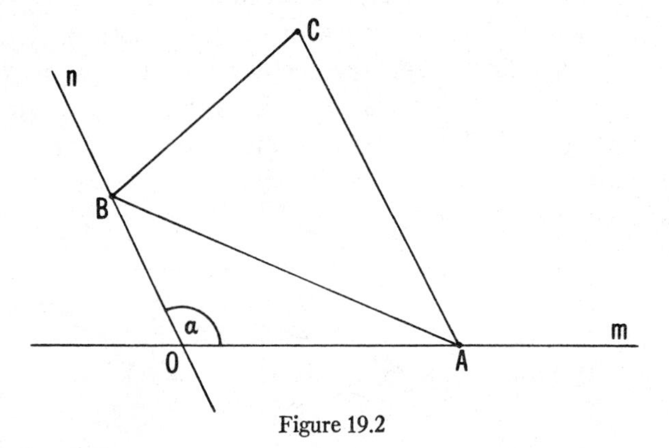

Figure 19.2

Now let us suppose that the vertices A and B of triangle ABC are constrained respectively to move along two lines, m and n, intersecting in an angle α (see Figure 19.2). We require the locus of the third vertex, C.

As B moves along n toward O, A gets pushed away from O, and the triangle AOB constantly changes. However, as this happens, the circumcircles for the triangles AOB are all the same size. In each case, the chord AB subtends the fixed angle α at the point O on the circumference. The length of AB never changes, and there is only one circle in which a chord of length AB can cut off a segment whose angle is equal to α. Of course, the position of this circle changes with the position of the chord AB.

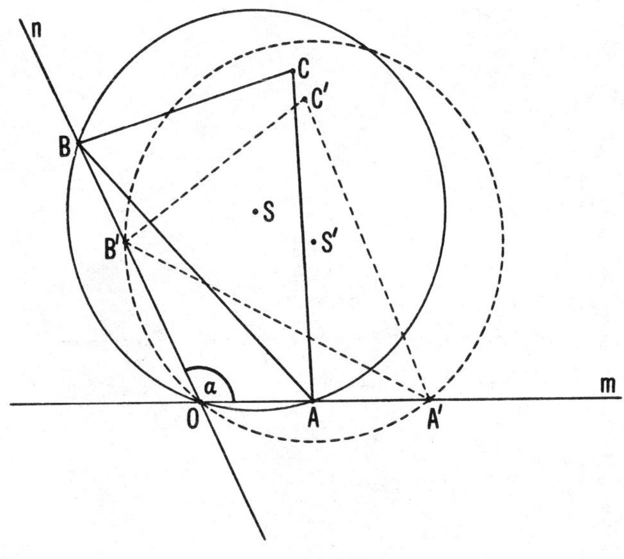

Figure 19.3

Let us, then, choose any particular position, e.g., the one drawn in solid lines in Figure 19.3, and construct the circumcircle of $\triangle AOB$. The circumcircle in any other position, e.g., the one drawn in dashed lines, is the same size, and since the length of AB is unchanged, the circle and the triangle ABC remain in the same relative position during the prescribed motion. It is as if the circle had been glued to the triangle and the combined figure moved. Consequently—and this is the point—the motions of A and B cause the attached triangle and circle to move so that the point O is always on the circumference of the circle. We might think of the figure as being moved so that the arc of the circle is "fed through" the point O while A and B slide on their lines m and n.

When B has passed through O, the chord AB subtends the supplement of α. But this is just the angle in the other segment of the attached circle. Hence the circle passes through O even as B (and later A) crosses O in its travels; see Figure 19.4.

To the moving circle with its attached triangle, we add the line joining the vertex C to the centre S of the circle, and we denote by U and V its intersections with the circle (see Figure 19.5). As the circle moves, the points A, B, U, V move with it; the arc AU, sub-

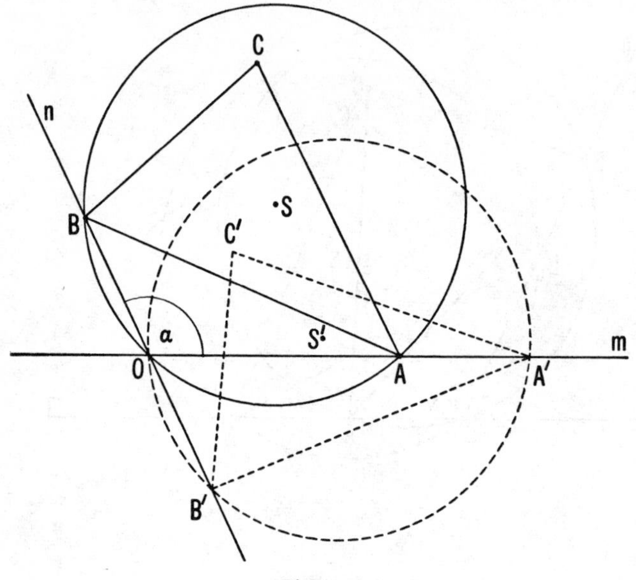

Figure 19.4

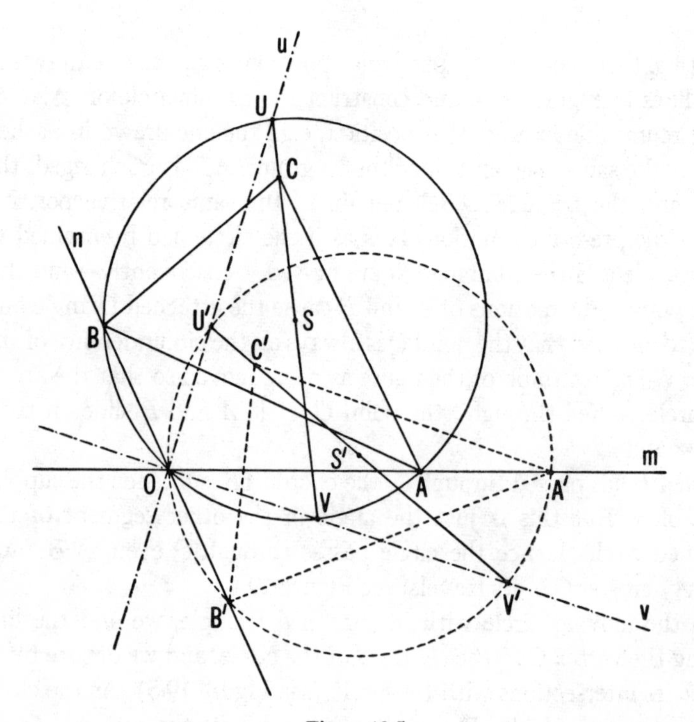

Figure 19.5

tended by $\measuredangle AOU$, moves without changing its length. Therefore, $\measuredangle AOU$ remains the same during this motion, and since its arm OA remains on the line m, its other arm connecting the fixed point O with the moving point U must remain fixed. Therefore, the point U moves along a line, say u, toward O. Similarly, V moves along a line v. As B and A run along their lines n and m, the points U and V run along the lines u and v.

Since UV is a diameter of the circle, $\measuredangle VOU$ is a right angle; that is, lines u and v are perpendicular. So, as A and B move along m and n, U and V move along a pair of perpendicular lines. The vertex C is on the line through U and V and hence, by our preliminary result, C moves along an ellipse with centre O and semi-axes along u and v of lengths CV, CU, respectively.

We conclude that, *when a triangle moves so that each of two of its vertices travels along a straight line, then the third vertex traces an ellipse.*

REFERENCE

H. Dorrie, *100 Great Problems of Elementary Mathematics*, Holt, Rinehart, and Winston, 1969, New York.

Solutions to Exercises

Essay 1

1. Let the times of arrival be x o'clock and y o'clock. Then x and y occur at random in the interval $[0, 1]$. Let P be the point in the unit square with coordinates (x, y). In order to have a meeting, $|x - y| \leq \frac{1}{4}$, i.e., $-\frac{1}{4} \leq x - y \leq \frac{1}{4}$. All points with coordinates x, y in the unit interval and satisfying the above inequalities $x - y \geq -\frac{1}{4}$ and $x - y \leq \frac{1}{4}$ are located in the shaded strip of Figure S1. Its area is $1 - 2(\triangle ABC) = 1 - (9/16) = 7/16$, which is equal to the required probability.

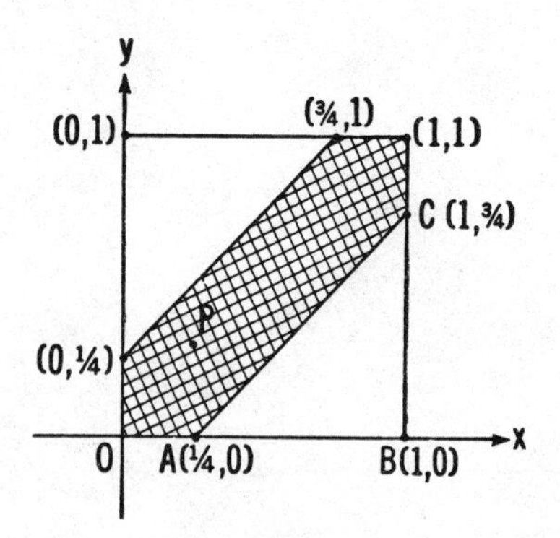

Figure S1

2. Let the length of the rod be taken as the unit length; let one end of the rod be the origin and let x and y be the coordinates of the breaks. Then x, y occur at random in $[0, 1]$, corresponding to a point $P(x, y)$ which has an equal chance of occurring anywhere in the unit square.

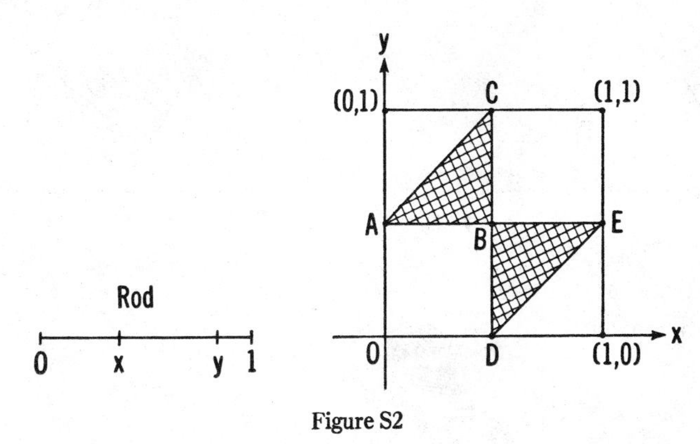

Figure S2

We require the probability that the triangle inequality is satisfied, i.e., that the sum of each two of the pieces exceeds the third. This is satisfied if each piece is less than one-half the length of the rod, and otherwise not. For definiteness, suppose $x \leq y$. Then we must have $x < \frac{1}{2}$, $y - x < \frac{1}{2}$, and $1 - y < \frac{1}{2}$. Each of these inequalities defines a region in the unit square (see Figure S2) and triangle ABC is the intersection of these regions. So P must be in $\triangle ABC$. Similarly, if $x \geq y$, P must lie in $\triangle BDE$. The required probability, then, is given by the sum of the areas of these triangles, which is easily seen to be $\frac{1}{4}$.

3. Let the circumference be taken as the unit of length. Since the circle can always be rotated about its centre to bring the vertex A into a pre-assigned position on the circumference (say the top), we may assume the point A to be fixed in this position. Cutting the circle at A and stretching it flat we obtain a unit segment $ABCA$ with points B and C chosen at random in it.

The requirement that $\triangle ABC$ be acute-angled means that each arc of the circle determined by the points A, B, and C must be less than one-half the circumference. This identifies Exercises 2 and 3, giving the answer $\frac{1}{4}$.

4. Let the equilateral triangle be $\triangle ABC$ with side s, and denote the lengths of PX, PY, PZ by x, y, z, respectively (see Figure S3).

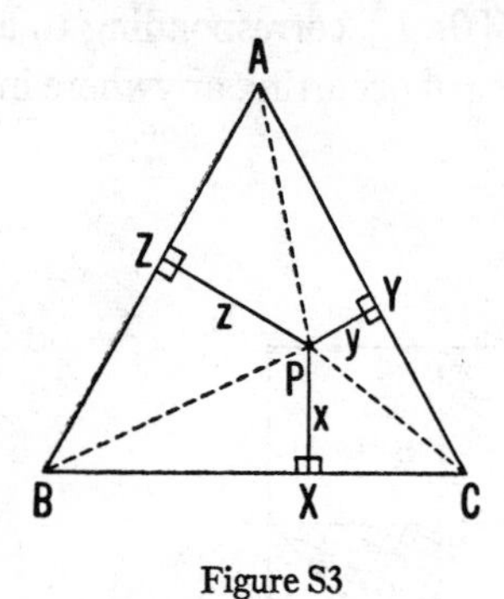

Figure S3

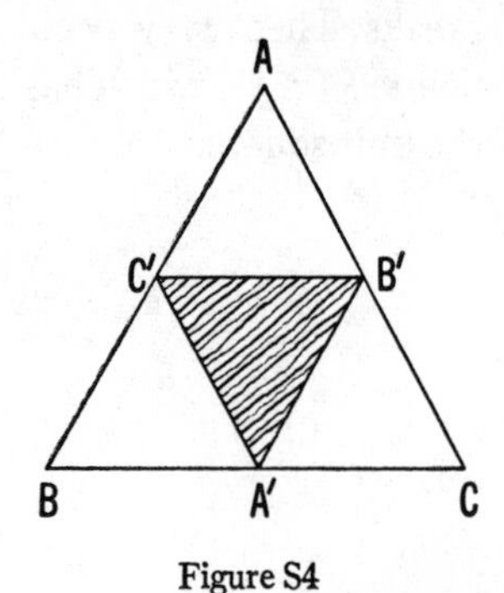

Figure S4

(i) Then

$$\begin{aligned}\triangle ABC &= \triangle PAB + \triangle PBC + \triangle PCA \\ &= \tfrac{1}{2}sz + \tfrac{1}{2}sx + \tfrac{1}{2}sy \\ &= \tfrac{1}{2}s(x + y + z).\end{aligned}$$

But

$$\triangle ABC = \tfrac{1}{2}sh,$$

where h is an altitude of $\triangle ABC$. This implies that for all positions of P,

$$x + y + z = h.$$

(ii) x, y, z represent the sides of a triangle if and only if each is less than $h/2$. This restricts P to the "medial triangle", whose vertices are the midpoints of the sides. The required probability, then, is the ratio $\triangle A'B'C'/\triangle ABC$, which is easily seen to be $1/4$. (See Figure S4.)

Note that Exercise 4 could also have been solved by the method used in solving Exercises 2 and 3. In those cases, we required that three numbers x, y and z satisfy the conditions

1) they are equally likely to lie anywhere between 0 and 1,
2) each is less than $1/2$,

and these conditions determined a region occupying precisely $1/4$ of the unit square.

The only modification in the present case is that our numbers x, y, z

1) lie at random in the interval $[0,\ h]$,
2) are each less than $h/2$.

The modified conditions determine a region occupying precisely $1/4$ of a square of side h, and the required probability is the ratio

$$\frac{\frac{1}{4}h^2}{h^2} = \frac{1}{4}.$$

Indeed, the special choice of unit length for the rod in Exercise 2 and the circumference of Exercise 3 was made for the sake of simplicity; any other constant would have led to the same answer.

Essay 4

1. Let A_s, A_d, A_m, B_s, B_d, $\cdots$ denote "Arnold is the shipper", "Arnold is the driver", and so on. We are given $C_s \Rightarrow B_d$, which we interpret

$$\sim C_s + B_d = 1.$$

Similarly we have

$$\sim C_d + B_m = 1, \qquad B_s + A_d = 1, \qquad \text{and} \qquad \sim A_m + C_d = 1.$$

Hence

$$(\sim C_s + B_d)(\sim C_d + B_m)(B_s + A_d)(\sim A_m + C_d) = 1.$$

This gives

$$\begin{aligned}&\sim C_s \sim C_d B_s \sim A_m + \sim C_s \sim C_d B_s C_d + \sim C_s \sim C_d A_d \sim A_m \\ &\quad + \sim C_s \sim C_d A_d C_d + \sim C_s B_m B_s \sim A_m + \sim C_s B_m B_s C_d \\ &\quad + \sim C_s B_m A_d \sim A_m + \sim C_s B_m A_d C_d + B_d \sim C_d B_s \sim A_m \\ &\quad + B_d \sim C_d B_s C_d + B_d \sim C_d A_d \sim A_m + B_d \sim C_d A_d C_d \\ &\quad + B_d B_m B_s \sim A_m + B_d B_m B_s C_d + B_d B_m A_d \sim A_m + B_d B_m A_d C_d = 1,\end{aligned}$$

which reduces, through obvious contradictions, to

$$\begin{aligned}&\sim C_s \sim C_d B_s \sim A_m + \sim C_s \sim C_d A_d \sim A_m + \sim C_s B_m A_d \sim A_m \\ &\qquad = \sim C_s \sim A_m (\sim C_d B_s + \sim C_d A_d + B_m A_d) = 1.\end{aligned}$$

This means

$$\sim C_s = 1, \quad \sim A_m = 1, \quad \text{and} \quad \sim C_d B_s + \sim C_d A_d + B_m A_d = 1.$$

Now, clearly $C_d + \sim C_d = 1$. Then

$$(C_d + \sim C_d)(\sim C_d B_s + \sim C_d A_d + B_m A_d) = 1,$$

i.e., $\sim C_d(B_s + A_d + B_m A_d) = 1$, which gives $\sim C_d = 1$ and $B_s + A_d + B_m A_d = 1$. From $\sim C_d = \sim C_s = 1$, it follows that $C_m = 1$. Consequently $B_m = 0$, and then $B_s + A_d = 1$.

But the shipper is either Brown or Arnold (since Clark is the manager), i.e., $B_s + A_s = 1$. Thus

$$(B_s + A_s)(B_s + A_d) = B_s(B_s + A_d) = 1.$$

Hence $B_s = 1$, and $A_d = 1$ follows. Thus Clark is the manager, Brown the shipper, and Arnold the driver.

2. Let A, B, C respectively denote "A is guilty", etc. Then we have $\sim A \Rightarrow BC$, i.e., $A + BC = 1$, and also $\sim A + B = 1$, and $\sim B + \sim C = 1$. Thus $(A + BC)(\sim A + B)(\sim B + \sim C) = 1$, which reduces to $AB \sim C = 1$, implying that C is the only innocent one.

3. If X makes the statement Y, then we have that
(i) X is telling the truth, and Y is true, or
(ii) X is lying, and Y is false.
That is, we have both $X \Rightarrow Y$ and $\sim X \Rightarrow \sim Y$; i.e., $\sim X + Y = 1$ and $X + \sim Y = 1$. Thus we have

$$\begin{array}{lll}
A \Rightarrow \sim B + \sim C, & \text{giving } \sim A + \sim B + \sim C = 1 & (1) \\
\sim A \Rightarrow \sim(\sim B + \sim C) \equiv BC, & A + BC = 1 & (2) \\
B \Rightarrow \sim A, & \sim B + \sim A = 1 & (3) \\
\sim B \Rightarrow A, & B + A = 1 & (4) \\
C \Rightarrow \sim A \sim B, & \sim C + \sim A \sim B = 1 & (5) \\
\sim C \Rightarrow \sim(\sim A \sim B) \equiv A + B, & C + A + B = 1 & (6).
\end{array}$$

Now (4) times (5) gives

$$\sim CB + \sim CA = \sim C(B + A) = 1, \quad \text{giving} \quad \sim C = 1.$$

This means $C = 0$, giving $A = 1$ in (2). From (3), then, $\sim B = 1$. Hence only A is telling the truth.

4. We have

$$
\begin{array}{lll}
A \Rightarrow BC, & \text{giving } \sim A + BC = 1 & (1) \\
\sim A \Rightarrow \sim(BC) \equiv \sim B + \sim C, & A + \sim B + \sim C = 1 & (2) \\
B \Rightarrow A, & \sim B + A = 1 & (3) \\
\sim B \Rightarrow \sim A, & B + \sim A = 1 & (4) \\
C \Rightarrow \sim A \sim B, & \sim C + \sim A \sim B = 1 & (5) \\
\sim C \Rightarrow \sim(\sim A \sim B) \equiv A + B, & C + A + B = 1 & (6).
\end{array}
$$

Now (3) times (4) gives $AB + \sim A \sim B = 1$, i.e., either A and B are both truth-tellers or they are both liars. If both are truth-tellers, then (1) gives $C = 1$, while (5) gives $\sim C = 1$, a contradiction. Hence $A = 0$ and $B = 0$. Now (6) gives $C = 1$. These latter values satisfy all six equations. Thus only C tells the truth.

5. Let H, J, D, $\cdots$ denote "Harry did it", "James did it", etc. The statements of the boys are, respectively, CG, DT, TC, HC, DJ. In four of the five compound statements, one statement is true and the other false; in the fifth both are false. Thus in all cases, the truth-value is 0; i.e.,

(1) $$CG = DT = TC = HC = DJ = 0.$$

For the same reason, four of the compound statements

$$C + G, \quad D + T, \quad T + C, \quad H + C, \quad D + J$$

have truth-value 1, while the fifth has value 0. Consequently,

(2) $$(C + G)(D + T)(T + C)(H + C)(D + J) = 1 \cdot 1 \cdot 1 \cdot 1 \cdot 0 = 0.$$

Using (1), this reduces simply to

(3) $$CDCCD \equiv CD = 0.$$

Since only one of $C+G$, $D+T$, $T+C$, $H+C$, $D+J$ is 0, then we have

$$\begin{aligned}&(C+G)(D+T)(T+C)(H+C)\\&\quad+(C+G)(D+T)(T+C)(D+J)\\&\quad+(C+G)(D+T)(H+C)(D+J)\\&\quad+(C+G)(T+C)(H+C)(D+J)\\&\quad+(D+T)(T+C)(H+C)(D+J)=1,\end{aligned}$$

where the left side contains the sum of the products of the compound statements taken four at a time. Using (1) and (3), this gives simply $CJ = 1$. Hence Charlie and James stole the apples.

6. Let A_r, A_w, A_b, B_r, $\cdots$ denote "A is red", etc. The given statements are A_r, $\sim B_r$, and $\sim C_b$. Now, some counter is red; hence

$$A_r + B_r + C_r = 1.$$

Similarly,

$$A_w + B_w + C_w = 1, \qquad A_b + B_b + C_b = 1.$$

Also, counter A has one of the three colours; hence

$$A_r + A_w + A_b = 1.$$

Similarly,

$$B_r + B_w + B_b = 1, \qquad C_r + C_w + C_b = 1.$$

Also, only one counter is of a given colour, so

$$A_rB_r = A_wB_w = A_bB_b = A_rC_r = \cdots = 0.$$

Finally, A can have only one colour, hence

$$A_rA_w = A_rA_b = A_wA_b = 0.$$

Similarly,

$$B_rB_w = B_rB_b = \cdots = C_rC_w = \cdots = 0.$$

Now then, we know that only one of the given statements is true. Hence

$$A_rB_rC_b + \sim A_r \sim B_rC_b + \sim A_rB_r \sim C_b = 1.$$

But $A_rB_r = 0$, giving

$$\sim A_r \sim B_rC_b + \sim A_rB_r \sim C_b = 1.$$

Multiplying by $A_r + B_r + C_r = 1$, we get just

$$\sim A_rB_r \sim C_b = 1.$$

Hence B is red, and C is not blue; but it is not red either (since B is). Hence C is white, and A blue.

7. Let A_a, A_c, A_f, A_d, C_a, $\cdots$ denote "Albert's last name is Albert", "Albert's last name is Charles", etc. Then we have

$$(1) \qquad A_a = C_c = F_f = D_d = 0$$

since no one has the same first and last names. Also, we have in particular that

$$(2) \qquad D_a = 0.$$

Let the Christian name of Mr. Frederick be X, and let the surname of Charles be y. Then

$$X_f = 1, \qquad C_y = 1,$$

and from the long complicated paragraph, we have that $Y_x = 1$. Then for some choice of X and Y among A, C, F and D (X and Y are different) we have

$$X_fY_xC_y = 1.$$

Substituting all possible values and summing gives an expression whose truth-value must be 1 (some choice is valid); i.e.,

$$A_fC_aC_c + A_fF_aC_f + A_fD_aC_d + \cdots = 1.$$

Working this out, and using (1) and (2), we get just $D_f A_d C_a = 1$. Hence Mr. Dick's first name is Albert.

Essay 6

Let n be chosen and let the multiples a, $2a$, $\cdots$, $(n-1)a$ be marked on a number-line. Also, let the same or another number-line have each interval between consecutive integers divided into n equal parts. (See Figure S5.)

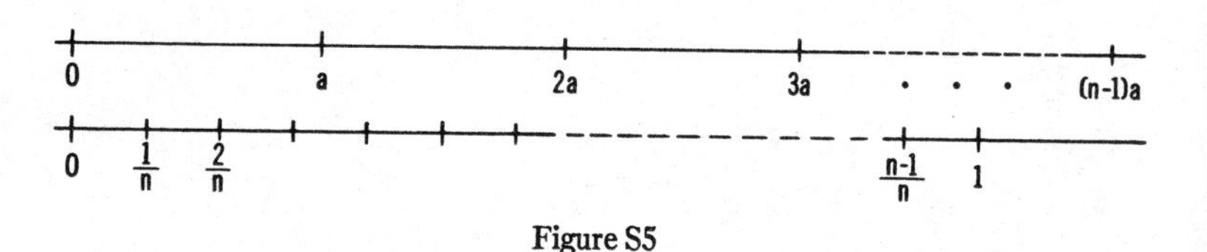

Figure S5

Now wrap both number-lines around a circle C of unit circumference, in the clockwise direction, beginning with the origin at the same point O in each case (see Figure S6). The divisions of the second line divide the circumference into n equal arcs. The multiples determine $n-1$ points on C. Now if one of these points occurs in either arc adjacent to the origin O, the required conclusion follows.

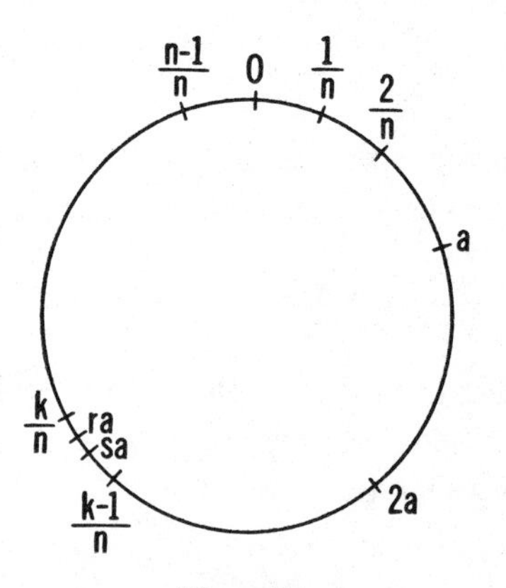

Figure S6

Otherwise, the $n-1$ points lie in the other $n-2$ arcs. It follows, by Dirichlet's Pigeon-hole Principle that at least two of them occur in the same arc. Suppose they are the multiples ra and sa $(r < s)$.

(Consider the arcs to be half-closed; this prevents both points from being endpoints.) This means that the difference $sa - ra$ differs from a whole number of circumferences, i.e., a whole number, by an amount less than $1/n$. That is, $(s - r)a$ differs from an integer by less than $1/n$. But $(s - r)a$ is, itself, one of the given multiples $a, 2a, \cdots, (n - 2)a$, because r and s are two different numbers from the set $\{1, 2, \cdots, n - 1\}$.

Essay 7

1. See Figure S7.

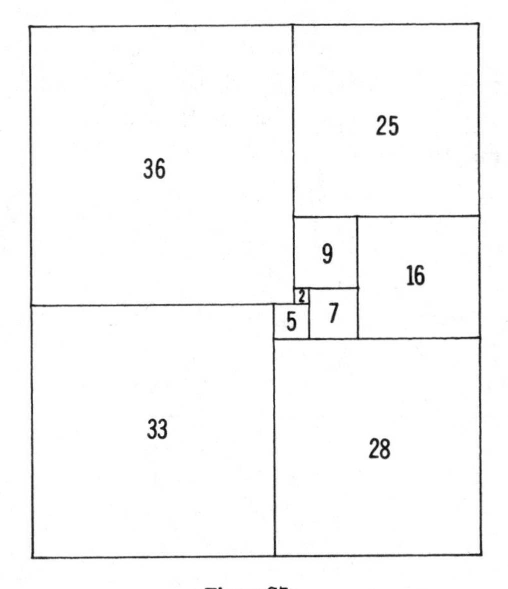

Figure S7

2. The 9-square tiling of the 32×33 rectangle given at the beginning of the essay can be extended to contain any number of squares by repeatedly constructing squares outwardly on the larger side (see Figure S8). No two squares are equal since no new square can duplicate an existing square. The same is true of the 9-square tiling of the 61×69 rectangle which is a solution of Exercise 1.

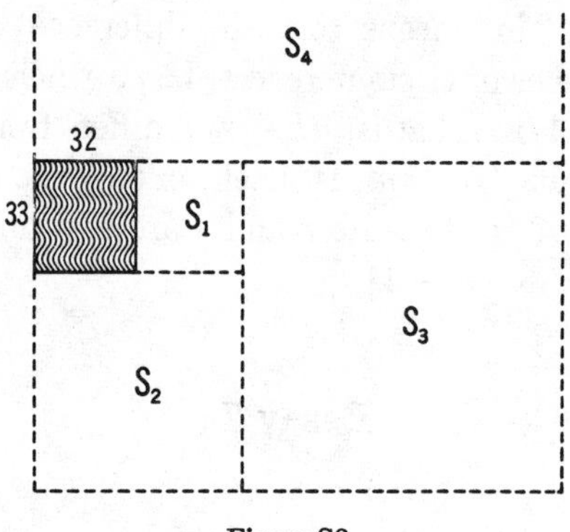

Figure S8

3. (a) If S were to have a side BC on the base, then not all triangles touching B or C can be larger than S without overlapping. [See Figure S9(a).] Thus S is "point down".

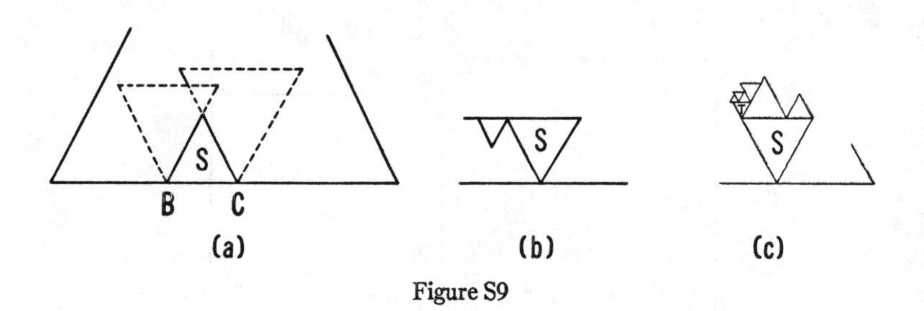

Figure S9

(b) The upper side of S cannot be continued beyond S lest S not be the smallest triangle touching the base. Hence the possibility of a single triangle in contact with the upper side of S cannot occur. Part (b) now follows as did part (a). [See Figure S9(b), (c).]

(c) No. The diagram shows that the single point of contact does not necessarily persist through the next stage (here T touches S at a vertex, which S could not do to the base).

Essay 9

1. Reflect P in each side of the angle α to obtain P' and P''. (See Figure S10.) We claim that the segment $P'P''$ cuts the arms at the required points Q and R. The perimeter of any other triangle $PQ'R'$ is

$$\begin{aligned} PQ' + Q'R' + R'P &= P'Q' + Q'R' + R'P'' \\ &\geq \text{the straight segment } P'P'', \end{aligned}$$

which is clearly the perimeter of $\triangle PQR$.

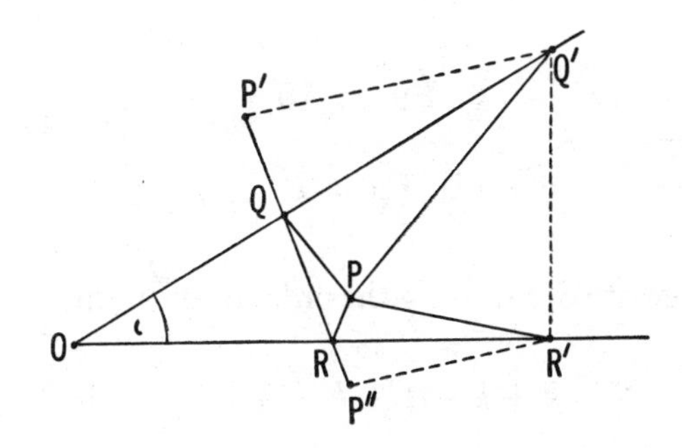

Figure S10

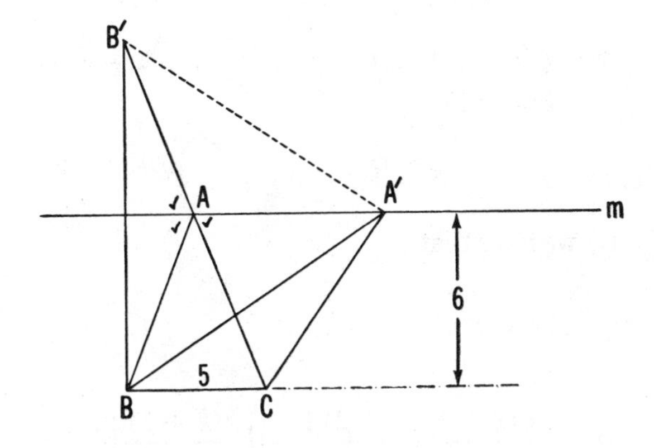

Figure S11

2. For base $BC = 5$ and area 15, the altitude to BC must be 6. Hence the third vertex A lies on a line m parallel to BC and 6 units away (see Figure S11). To locate A on m so that $\triangle ABC$

has minimum perimeter, reflect B in m to give B'. Let $B'C$ cut m at A. Then for any other point A' on m, the perimeter of $\triangle A'BC$ is

$$\begin{aligned} BC + A'B + A'C &= BC + A'B' + A'C \\ &\geq BC + \text{the straight segment } B'C \\ &= \text{the perimeter of } \triangle ABC. \end{aligned}$$

It is not difficult to show that $AB = AC$. Hence each side is

$$\sqrt{(5/2)^2 + 6^2} = 13/2.$$

Essay 10

1. (i) Let $$S = 1 + 2 + 3 + \cdots + k.$$

We rewrite S, reversing the order of the terms:

$$S = k + k - 1 + k - 2 + \cdots + 1.$$

Adding gives

$$\begin{aligned} 2S &= k + 1 + k + 1 + k + 1 + \cdots + k + 1 \\ &= k(k+1). \end{aligned}$$

Hence $$S = k(k+1)/2.$$

(ii) By (i) we have that

$$[\sum_1^{k+1} i]^2 - [\sum_1^{k} i]^2$$

$$= \left[\frac{(k+1)(k+2)}{2}\right]^2 - \left[\frac{k(k+1)}{2}\right]^2$$

$$= \left(\frac{k+1}{2}\right)^2 [k^2 + 4k + 4 - k^2] = (k+1)^2(k+1)$$

$$= (k+1)^3.$$

(iii) If

$$\sum_{1}^{k} i^3 = [\sum_{1}^{k} i]^2$$

then

$$\sum_{1}^{k+1} i^3 = \sum_{1}^{k} i^3 + (k+1)^3 = [\sum_{1}^{k} i]^2 + (k+1)^3$$

$$= [\sum_{1}^{k+1} i]^2, \qquad \text{by (ii).}$$

(iv) For $k = 1$, each side of identity (1) has the value 1.

2. There are $C(10, 4) = 210$ numbers with 4 different digits, $10C(9, 2) = 360$ numbers with digits 2 alike and 2 different, $C(10, 2) = 45$ numbers with digits 2 alike and 2 alike, $9C(10, 1) = 90$ numbers with digits 3 alike, giving a total of 705.

3. (a) $D_1 = 2088$; $E_1 = 8082$.
Direct calculation and inspection establishes parts (b) and (c).

7. (a) The n-th number in the middle row has n^2 at the bottom of its column. The number of integers in this column is $n^2 - (n-1)^2 = 2n - 1$, which is always odd. The middle one in the column, then, is $(\frac{1}{2}[(2n-1)+1])$-th from the top (i.e., n-th from the top). Since it is in the n-th place past the bottom number of the previous column, which is $(n-1)^2$, its value is $(n-1)^2 + n = n^2 - n + 1$.

(b) The number 3 and every third number thereafter is in position $2 + 3k$ for $k = 0, 1, 2, \cdots$. By part (a), its value therefore is

$$(2+3k)^2 - (2+3k) + 1 = 9k^2 + 9k + 3;$$

and every number of this form is clearly divisible by 3. The solutions of parts (c) and (d) are similar.

(e) Consecutive numbers in the n-th and $(n+1)$-th positions of the middle row have the forms $n^2 - n + 1$ and $(n+1)^2 - (n+1) + 1 = n^2 + n + 1$. Their product is $(n^2+1)^2 - n^2 = (n^2+1)^2 - (n^2+1) + 1$, implying that it occurs in the (n^2+1)-th place in the row, as required. The square at the bottom of the column containing the smaller factor, $n^2 - n + 1$, is n^2.

8. The entry in position (n, n) is the sum $1 + 5 + 9 + \cdots$ of n terms and is equal to $n[2 + (n-1)4]/2 = 2n^2 - n$. The common difference for column n is the n-th odd number, $2n - 1$. Hence the number in position (m, n) $(m \geq n)$ is

$$[\text{entry in } (n,n)] + [(m-n) \times \text{common difference of } n\text{-th column}]$$
$$= 2n^2 - n + (m-n)(2n-1) = (2n-1)m.$$

The sum of the numbers in row k, then, is the sum of the numbers in positions $(k, 1), (k, 2), \cdots, (k, k)$, which is

$$\begin{aligned}(2\cdot1-1)k + (2\cdot2-1)k + (2\cdot3-1)k + \cdots + (2\cdot k-1)k \\ = k[2(1+2+\cdots+k) - k] \\ = k[k(k+1) - k] \\ = k^3.\end{aligned}$$

Essay 11

1. The angle in question is the sum of an angle x of elevation and an angle y of depression, each of which increases continuously to a right angle as the observer approaches the statue. (See Figure S12.)

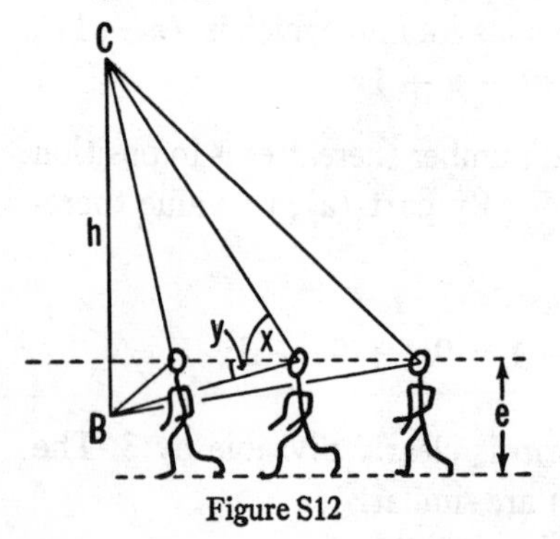

Figure S12

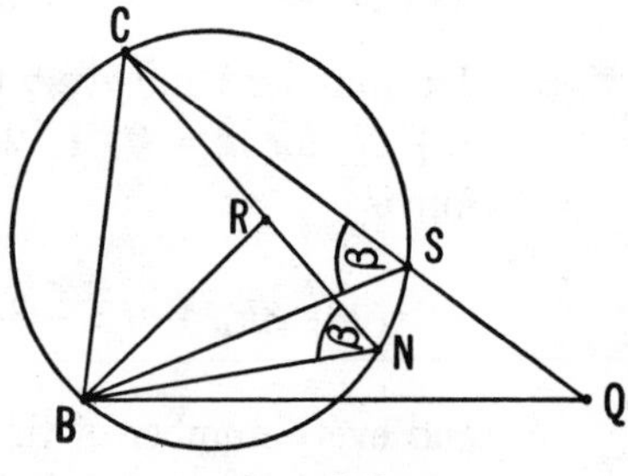

Figure S13

2. Use the hint and the fact that an exterior angle of a triangle exceeds either non-adjacent interior angle. (See Figure S13.)

3. (i) The line $y = \frac{1}{2}(b + c)$, the perpendicular bisector of the chord BC, must pass through the centre Z whose ordinate therefore is $\frac{1}{2}(b + c)$. The abscissa of the centre is equal to OM', and is given by (see Figure 11.4)

$$(OM')^2 = OB \cdot OC \quad (OM' \text{ is a tangent and } OBC \text{ a secant});$$

i.e., the abscissa $= OM' = \sqrt{bc}$. Since OM' is a tangent, the radius ZM' is vertical, so its length is $\frac{1}{2}(b + c)$.

(ii) Let the midpoint of BC be T. We saw in (i) that $\sqrt{bc} = OM' = ZT$, while $\frac{1}{2}(b + c) = \text{radius } ZM' = ZB$. Thus $\sqrt{bc}$ is a leg and $\frac{1}{2}(b + c)$ is the hypotenuse of right-triangle ZTB, implying the required inequality.

(iii) Since $(b - c)^2 \geq 0$, $(b - c)^2 + 4bc \geq 4bc$; hence $(b + c)^2 \geq 4bc$, $b + c \geq 2\sqrt{bc}$ or $\frac{1}{2}(b + c) \geq \sqrt{bc}$. Here equality holds only if $b = c$.

4. From the given data $b = p - e = 8'$, $c = p + h - e = 18'$, and the required distance is

$$\sqrt{bc} = \sqrt{8 \cdot 18} = 12 \text{ feet.}$$

Essay 12

1. Let $f(x_0) = n$. Now $f^*(n) = \max x$ such that $f(x) \leq n$, i.e., such that $f(x) \leq f(x_0)$. Clearly x_0 is one value of x satisfying this inequality. Since f is strictly increasing, x_0 is the largest such x. Thus $f^*(n) = f^*(f(x_0)) = x_0$ for arbitrary x_0.

2. Let $F(n)$, $G(n)$ denote the sequences of triangular numbers and non-triangular numbers, respectively. Then

$$f(n) = F(n) - n = \frac{n(n+1)}{2} - n = \frac{n(n-1)}{2},$$

and

$$f^*(n) = \text{the number of positive } m \text{ such that } f(m) < n$$

$$= \text{the number of positive } m \text{ such that } \frac{m(m-1)}{2} < n,$$

$$\text{i.e., such that } m(m-1) < 2n.$$

Now just as in our treatment of the non-square numbers (see §III, pp. 97–8) the inequality $m(m-1) < n$ led to the formula $G(n) = n + \langle\sqrt{n}\rangle$, so our present inequality

$$m(m-1) < 2n$$

leads to the formula $G(n) = n + \langle\sqrt{2n}\rangle$.

3. (a) Suppose $k \leq \sqrt{n} < k + 1/2$. Then $[\sqrt{n}] = k$, and also $\langle\sqrt{n}\rangle = k$. Now

$$k^2 \leq n < k^2 + k + 1/4$$

and, adding k to each member, we obtain

$$k^2 + k \leq n + [\sqrt{n}] < k^2 + 2k + 1/4,$$

whence

$$k^2 < n + [\sqrt{n}] < k^2 + 2k + 1,$$

implying

$$k < \sqrt{n + [\sqrt{n}]} < k + 1$$

and

$$[\sqrt{n + [\sqrt{n}]}] = k = \langle\sqrt{n}\rangle.$$

Part (b) is similar.

4. Let T be the n-th positive integer not of the form $[e^k]$, k a positive integer, and suppose that it occurs between $[e^m]$ and $[e^{m+1}]$. (See Figure S14.)

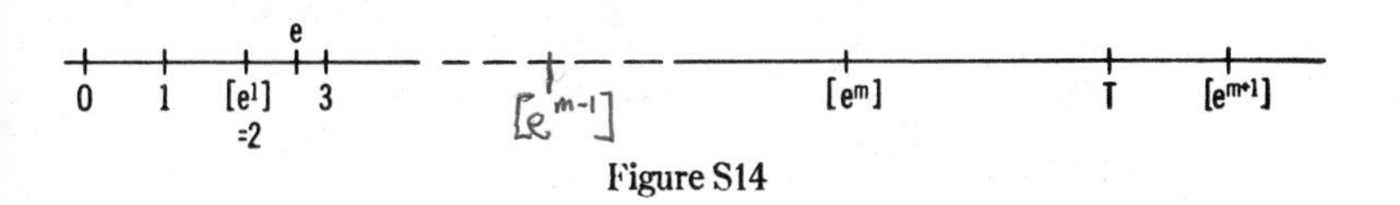

Figure S14

Then

$$T - m = n, \text{ or } T = n + m.$$

Since e is a transcendental number (i.e., *not* a root of a polynomial equation with integer coefficients), e^k is *not* an integer for any positive integer k; for, if e^k = an integer a, then $e^k - a = 0$ so that e would be a zero of the polynomial $x^k - a = 0$. It follows that $[e^k] < e^k$ for all k. Also, since T is not of the form $[e^k]$ and lies between $[e^m]$ and $[e^{m+1}]$, we see that $T \leq [e^{m+1}] - 1$. (See Figure S15.) Therefore

$$e^m < [e^m] + 1 \leq T \leq [e^{m+1}] - 1$$

and

$$e^m < T + 1 \leq [e^{m+1}] < e^{m+1}.$$

Taking logarithms of the first, second and last members, we obtain

$$m < \log (T + 1) < m + 1$$

and conclude

$$[\log (T + 1)] = [\log (n + 1 + m)] = m.$$

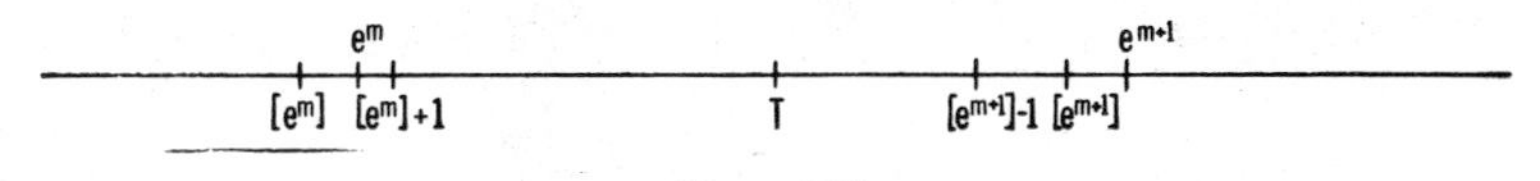

Figure S15

Using this repeatedly we get

$$\begin{aligned} T &= n + m \\ &= n + [\log (n + 1 + m)] \\ &= n + [\log (n + 1 + [\log (n + 1 + m)])] \\ &= \cdots \end{aligned}$$

as far as one pleases. We need to show that

$$Z = n + [\log (n + 1 + [\log (n + 1)])]$$

is the constant value of all these expressions.

First we show that n exceeds $[e^{m-1}]$. For, if not,

$$n \leq [e^{m-1}] < e^{m-1}$$

would yield

$$T = n + m \leq [e^{m-1}] + m,$$

and

$$e^m < T \leq [e^{m-1}] + m < e^{m-1} + m;$$

that is,

$$e^m < e^{m-1} + m.$$

This leads to

$$e^{m-1}(e - 1) < m$$

and, since $e - 1 > 1$, to

$$e^{m-1} < m.$$

But

$$e^{m-1} = 1 + (m - 1) + \frac{(m - 1)^2}{2!} + \cdots \geq 1 + (m - 1) = m,$$

a contradiction.

Then n occurs between $[e^{m-1}]$ and T. This interval also contains $[e^m]$. We consider the subintervals $[[e^m], T)$ and $([e^{m-1}], [e^m])$ separately.

(a) If $[e^m] \le n < T$, then

$$e^m < [e^m] + 1 \le n + 1 \le T < [e^{m+1}] < e^{m+1},$$

giving

$$m < \log (n+1) < m+1, \qquad \text{and} \qquad [\log (n+1)] = m.$$

In this case

$$Z = n + [\log (n + 1 + m)] = T$$

as required.

(b) If $[e^{m-1}] < n < [e^m]$, then

$$e^{m-1} < [e^{m-1}] + 1 < n + 1 \le [e^m] < e^m,$$

so that

$$m - 1 < \log (n+1) < m \quad \text{and} \quad [\log (n+1)] = m - 1.$$

Thus

$$\begin{aligned} Z &= n + [\log (n + 1 + m - 1)] \\ &= n + [\log (n + m)] \\ &= n + [\log T]. \end{aligned}$$

Since $e^m < T < e^{m+1}$, we have $[\log T] = m$, so that

$$Z = n + m = T,$$

as required.

5. Obviously $(0, 0)$, $(1, 0)$, $(0, 1)$, $(2, 0)$, $(0, 2)$, $(1, 1)$ are all either winning positions or losing ones.

Suppose the assertion holds for (x, y) if $x + y < n$. Consider (X, Y), where $X + Y = n$.

Now, in general, there are many positions (x, y) which can be moved into from (X, Y). Imagine, then, that you receive the position (X, Y) from your opponent and that you move to (x, y). For every possible (x, y) that you might choose, $x + y < n$, implying that each is either a winner or a loser (for you, that is). Now if any of these is a winner for you, then (X, Y) is a loser for your opponent, because you will, of course, choose to move into that position (x, y) that wins for you. In this case (X, Y) is a loser because it loses for the player who moves into it (i.e. passes it along to his opponent). On the other hand, if none of the (x, y) are winners for you, then (X, Y) is a winner for your opponent, because you can't win no matter what you do. Thus from your opponent's point of view, and this is the point of view from which to appraise (X, Y), we see that (X, Y) is also either a winner or a loser.

Essay 13

1. We use the formula for k-gonal numbers derived on pp. 117–119 and find

$$
\begin{aligned}
p_n^{(k-1)} + p_{n-1}^{(3)} &= \frac{n}{2}[n(k-3) - k + 5] + \frac{n-1}{2}[(n-1)+1] \\
&= \frac{n}{2}[n(k-3) - k + 5 + n - 1] \\
&= \frac{n}{2}[n(k-2) - k + 4] \\
&= p_n^{(k)}.
\end{aligned}
$$

2. $x = 2mn$, where one of m, n is even. This implies x is divisible by 4.

3. If neither m nor n is divisible by 3, we have m and n of the form $3k + 1$ or $3k + 2$. In either case, m^2 and n^2 are of the form $3k + 1$, so that $y = m^2 - n^2$ is divisible by 3.

4. Suppose neither m nor n is divisible by 5. Then m, n are congruent to $1, -1, 2,$ or $-2 \pmod 5$, giving m^2, n^2 congruent to 1 or 4. If m^2 and n^2 are both congruent to the same number, then $y = m^2 - n^2$ is congruent to 0; otherwise $z = m^2 + n^2$ is congruent to 0.

5. Suppose q is not a prime, but equal to ab, where a, b are integers greater than 1. Then $2^q - 1 = (2^a)^b - 1$, which has the divisor $2^a - 1$. And $2^a - 1$ is not 1, lest $a = 1$; nor is it $2^q - 1$, lest $b = 1$. Hence $2^q - 1$ is composite; contradiction.

6. An even perfect number m is of the form $2^{n-1}(2^n - 1)$, where $2^n - 1$ is a prime (see the theorem on page 115). By the previous problem, n is a prime. Consequently $n = 2$, or n is of the form $4k + 1$ or $4k + 3$. (It is either even, or it is an odd prime).

   If $n = 2$, we get $m = 6$, which is acceptable.

   If $n = 4k + 1$,

$$m = 2^{4k}(2^{4k+1} - 1) = 16^k(2 \cdot 16^k - 1).$$

16^k ends in 6 for all k; thus $2 \cdot 16^k - 1$ ends in 1, and thus m ends in 6.

   If $n = 4k + 3$,

$$m = 2^{4k+2}(8 \cdot 16^k - 1) = 4 \cdot 16^k(8 \cdot 16^k - 1).$$

Since 16^k ends in 6, $4 \cdot 16^k$ ends in 4, $8 \cdot 16^k - 1$ ends in 7, and their product m ends in 8.

Essay 16

1. (a) If $k = l$, d is a power of 10 so that n/d terminates. If $k > l$, then

$$\frac{n}{d} = \frac{n \cdot 5^{k-l}}{d \cdot 5^{k-l}} = \frac{D}{2^k 5^k} = \frac{D}{10^k},$$

giving a terminating decimal. If $k < l$, multiply numerator and denominator by 2^{l-k} to obtain an equivalent fraction with 10^l in the denominator.

(b) Let the prime decomposition of b be $b = p_1^{\alpha_1}p_2^{\alpha_2} \cdots p_t^{\alpha_t}$. Then, if $d = p_1^{\beta_1}p_2^{\beta_2} \cdots p_t^{\beta_t}$, where the exponents are non-negative integers, the "b-mal" expansion of n/d terminates.

Proof: Let q be a positive integer such that each of $\alpha_1 q, \alpha_2 q, \cdots, \alpha_t q$ exceeds the maximum of $\beta_1, \beta_2, \cdots, \beta_t$.

$$d \cdot p_1^{\alpha_1 q - \beta_1} p_2^{\alpha_2 q - \beta_2} \cdots p_t^{\alpha_t q - \beta_t} = b^q,$$

so that the equivalent fraction is

$$\frac{n}{d} = \frac{n \cdot p_1^{\alpha_1 q - \beta_1} p_2^{\alpha_2 q - \beta_2} \cdots p_t^{\alpha_t q - \beta_t}}{b^q}$$

and evidently terminates.

2. We are concerned only with the part of the division process which provides digits in the quotient after the decimal point. After the decimal point, only zeros are "brought down" to usher in successive steps. The only possible remainders are $0, 1, 2, \cdots, d-1$, and if the process does not end, the remainder 0 cannot occur. In at most d steps, then, one of the remainders must repeat (Dirichlet's principle again). After the second appearance of this remainder, the process continues exactly as it did after its first occurrence (again a zero is brought down, etc.). Thus the process repeats over and over with a period of length not exceeding $d - 1$.

3. (a) This is simply a matter of verifying the three properties required in the definition.

(b) Employing the hint we get

$$f_1g_2 - g_1g_2 = g_2k_1p$$

and

$$f_1f_2 - f_1g_2 = f_1k_2p$$

whose sum is

$$f_1 f_2 - g_1 g_2 = (g_2 k_1 + f_1 k_2) p.$$

This implies the desired conclusion.

(c) We use the method of induction. Suppose the assertion holds for $m = k$. Now consider $k + 1$ pairs

$$f_i \equiv g_i \pmod{p}, \qquad i = 1, 2, \cdots, k + 1.$$

By assumption,

$$f_1 f_2 \cdots f_k \equiv g_1 g_2 \cdots g_k \pmod{p}$$

and

$$f_{k+1} \equiv g_{k+1} \pmod{p}.$$

These congruences, together with the result of (b), yield

$$f_1 f_2 \cdots f_{k+1} \equiv g_1 g_2 \cdots g_{k+1} \pmod{p}.$$

Thus the assertion holds for $m = k + 1$. Since it holds for $m = 2$, by part (b), it holds for all $m = 2, 3, 4, \cdots$.

4. The statement and proof parallel the details given in the essay with 10 replaced by b, 9 by $b - 1$, and the condition p is not 2 or 5 by p does not divide b. The working out of the substitutions is left to the reader.

5. This problem may be attacked either by generalizing the discussion on pp. 150–153 from a prime p to a number q relatively prime to 10, or by examining the division algorithm. We present the latter approach. It suffices to consider the case $r < q$ (see page 150).

For its decimal expansion to terminate, r/q must be expressible in the form $s/10^n$ for some integers s and n; i.e., there must be an integer m such that $mq = 10^n$. But this equation cannot hold if q is relatively prime to 10. Hence the expansion of r/q does not terminate; it is periodic (see Exercise 2).

In the long division process repeating remainders cause repeating digits in the quotient. Suppose equal remainders, r_m and r_n, occur in the m-th and n-th steps (places after the decimal point). Since every remainder except the first results from subtracting a multiple of the divisor from 10 times the previous remainder, we have

$$10r_{m-1} = aq + r_m$$

and

$$10r_{n-1} = bq + r_n.$$

The difference of these equations is

$$10(r_{m-1} - r_{n-1}) = q(a - b),$$

and since q is relatively prime to 10, q divides $r_{m-1} - r_{n-1}$. However, each remainder is a positive integer less than q, so the difference $r_{m-1} - r_{n-1}$ is less than q and hence can be divisible by q only if it is zero, i.e., if $r_{m-1} = r_{n-1}$. In other words, if $r_m = r_n$, then $r_{m-1} = r_{n-1}$, and so on back to the very first decimal place where r_1, the first remainder, starts the period.

6. This well known result is called Euler's generalization of Fermat's theorem, and its proof can be found in many books on number theory, e.g., Hardy and Wright, *Introduction to Number Theory*, or H. N. Wright, *Theory of Numbers* (First Course).

Bibliography

W. W. R. Ball, *Mathematical Recreations and Essays* (11th ed.), MacMillan, 1939, London.

E. T. Bell, *Men of Mathematics*, Simon and Schuster, 1961, New York.

R. Courant and H. Robbins, *What is Mathematics?*, Oxford University Press, 1941, New York.

N. A. Court, *College Geometry* (2nd ed.), Barnes and Noble, 1952, New York.

H. S. M. Coxeter, *Introduction to Geometry*, Wiley, 1961, New York.

H. Dorrie, *100 Great Problems of Elementary Mathematics*, Dover, 1958, New York.

H. Eves, *Introduction to the History of Mathematics*, Holt, Rinehart, and Winston, 1969, New York.

G. Hardy and E. Wright, *Introduction to the Theory of Numbers*, Clarendon, 1960, Oxford.

D. Hilbert and S. Cohn-Vossen, *Geometry and the Imagination*, Chelsea, 1952, New York.

L. Lyusternik, *Convex Figures and Polyhedra*, Dover, 1963, New York.

E. A. Maxwell, *Geometry for Advanced Pupils*, Oxford Press, 1949, Oxford.

NCTM Yearbook 28: *Enrichment Mathematics for High Schools*, 1963, Washington.

D. Pedoe, *The Gentle Art of Mathematics*, Penguin reprint, 1969, New York.

G. Pólya, *Mathematics and Plausible Reasoning*, Vol. 1, Princeton, 1954, Princeton.

H. Rademacher, *Lectures in Elementary Number Theory*, Blaisdell, 1964, New York.

H. Rademacher and O. Toeplitz, *The Enjoyment of Mathematics*, Princeton University Press, 1957, Princeton.

H. Steinhaus, *100 Problems in Elementary Mathematics*, Basic Books, 1964, New York.

H. Tietze, *Famous Problems of Mathematics*, Graylock, 1965, New York.

H. N. Wright, *First Course in Theory of Numbers*, Wiley, 1939, New York.

I. M. Yaglom and V. G. Boltyanskii, *Convex Figures*, Holt, Rinehart and Winston, 1961, New York.

A. M. Yaglom and I. M. Yaglom, *Challenging Mathematical Problems with Elementary Solutions*, Holden Day, 1964, San Francisco.

Index